Suzy Koontz's
MULTIPLY WITH ME!

Learning to MULTIPLY Can be Fun!

MATHMADEFUN
A Division of ACTS, Inc.

Math Made Fun
PO Box 4017
Ithaca, NY 14850
www.mathmadefun.com

This book is dedicated to my 2×2 daughters and my loving, supportive husband.

Acknowledgments

I am sincerely grateful to my editor Linda Glaser. Her numerous suggestions for restructuring the manuscript proved to be invaluable. I truly appreciate Laura Gates-Lupton for her many helpful ideas for improving the readability of the manuscript, and for providing copy editing. Thank you also to Mom, Dad, Bob, and Margaret for offering suggestions, proofreading and support. In addition, I appreciate the critique provided by the members of my critique group. Thanks to Laura Gates-Lupton, Jodie Mangor, Cathleen Banford, Johanna Husband, Andrea Hazard, and Sigrid Mortensen. Annie Zygarowicz also deserves gigantic thanks for doing an amazing job designing the cover and completing the interior graphics.

A six-year-old named Sarah deserves a special thanks for serving as my educational consultant. Her priceless grimaces, at key moments, showed me where to alter the text. Special thanks go to Sarah's three older sisters Lizzi, Emily, and Jessica, who at one time or another survived being my guinea pigs while I tested my ideas. I am so grateful to my four daughters and my husband Bruce for their love, patience, and understanding.

I appreciate the teachers and individuals who reviewed the book, wrote letters of recommendation or provided encouragement. Thanks to Ellen Mitterer, Renee Qamar, Lisa Neville, Rick Junge, Margaret Steinacher, Terri Stoff, Tracy Kirkman, Pam Wooster, Mary Psiaki, Jane Bruce-Robertson, Jacquie Lopez, Melody B., Pamela W., Nancy Reddy, Maria Rider, Nancy Saltzman, Bec Groves-Haley, Lisa Sanflippo, Karen Grace-Martin, Susan Hubbard, and Jessica Mitchell. Thank you also to the many other teachers and families who reviewed and tested the book.

Photo Credit: Jacqueline Conderacci © 2008

Math Made Fun
PO Box 4017
Ithaca, NY, 14850

www.mathmadefun.com

Lesson 1

Color every 2nd square using a light-colored highlighter, crayon or marker. These numbers will be 2, 4, 6, 8, 10, 12, 14, 16, 18, 20, 22, and 24.

1	2	3	4	5	6	7	8	9	10
11	12	13	14	15	16	17	18	19	20
21	22	23	24	25	26	27	28	29	30

1. Are all the even numbers colored?   Circle the correct answer: yes   no

2. Read the colored numbers.

3. Whisper the colored numbers.

4. How loud can you yell?   Circle the correct answer:

   Loud   Very Loud   Very, Very Loud   So Loud I Will Have to Cover My Ears

5. Hold your hands over your ears and shout the colored numbers.

6. Answer these math problems:

$$0+2= \underline{\quad} \qquad\qquad 2+2= \underline{\quad}$$

$$4+2= \underline{\quad} \qquad\qquad 6+2= \underline{\quad}$$

Lesson 2

1. Color every 2nd square using a light-colored highlighter, crayon or marker.

1	2	3	4	5	6	7	8	9	10
11	12	13	14	15	16	17	18	19	20
21	22	23	24	25	26	27	28	29	30

2. Are all the even numbers colored? Circle the answer: yes   no

3. Read the colored numbers.

4. Whisper the colored numbers.

5. Shout the colored numbers.   (Hold your hands over your ears!)

6. Would you like a double scoop of ice cream?   Circle the answer: Yes   No

7. Does double mean two scoops of ice cream?   Circle the answer: Yes   No

Underline the math symbols that mean double: (Warning! Some of these are tricky.)

Example: The double of the number 3 is	*3+3*	*2 threes*	*2×3*	*3×2*
8. The double of the number 2 is	2+2	2 twos	2×2	2×2
9. The double of the number 4 is	4+4+4	2 fours	2×4	4×2
10. The double of the number 5 is	5+5	2 fives	2×5	5×2
11. The double of the number 6 is	6+6	5 sixes	2×6	6×2
12. The double of the number 7 is	7+7	2 sevens	4×7	7×2
13. The double of the number 8 is	8+8+8	2 eights	2×8	8×2
14. The double of the number 9 is	9+9	2 nines	2×9	9×3

Now add on 2.

$0+2=$ ___

$2+2=$ ___

$4+2=$ ___

$6+2=$ ___

$8+2=$ ___

$10+2=$ ___

$12+2=$ ___

Lesson 3

1. Color every 2nd square using a light-colored highlighter, crayon or marker.

1	2	3	4	5	6	7	8	9	10
11	12	13	14	15	16	17	18	19	20
21	22	23	24	25	26	27	28	29	30

2. Read the colored numbers.

4. Whisper the colored numbers.

5. Shout the colored numbers.   (Hold your hands over your ears!)

6. Solve these problems:

$3+3=$ _____ $2 \times 3 =$ _____

$4+4=$ _____ $2 \times 4 =$ _____

$5+5=$ _____ $2 \times 5 =$ _____

6+6=_____   2×6 = _____

7+7=_____   2×7 = _____

8+8=_____   2×8 = _____

9+9=_____   2×9 = _____

10+10=_____   2×10 = _____

11+11=_____   2×11 = _____

12+12=_____   2×12 = _____

13+13=_____   2×13 = _____

Lesson 4

1. Color every 2nd square using a light-colored highlighter, crayon or marker.

1	2	3	4	5	6	7	8	9	10
11	12	13	14	15	16	17	18	19	20
21	22	23	24	25	26	27	28	29	30

2. Read, whisper and shout (really loud) the colored numbers.

3. Solve these problems:

$$\begin{array}{r} 1 \\ \times\,2 \\ \hline \end{array} \qquad \begin{array}{r} 2 \\ \times\,2 \\ \hline \end{array} \qquad \begin{array}{r} 3 \\ \times\,2 \\ \hline \end{array} \qquad \begin{array}{r} 4 \\ \times\,2 \\ \hline \end{array}$$

$$\begin{array}{r} 5 \\ \times\,2 \\ \hline \end{array} \qquad \begin{array}{r} 6 \\ \times\,2 \\ \hline \end{array} \qquad \begin{array}{r} 7 \\ \times\,2 \\ \hline \end{array} \qquad \begin{array}{r} 8 \\ \times\,2 \\ \hline \end{array}$$

$$\begin{array}{r} 2 \\ \times\,1 \\ \hline \end{array} \qquad \begin{array}{r} 2 \\ \times\,2 \\ \hline \end{array} \qquad \begin{array}{r} 2 \\ \times\,3 \\ \hline \end{array} \qquad \begin{array}{r} 2 \\ \times\,4 \\ \hline \end{array}$$

$$\begin{array}{r} 2 \\ \times\,5 \\ \hline \end{array} \qquad \begin{array}{r} 2 \\ \times\,6 \\ \hline \end{array} \qquad \begin{array}{r} 2 \\ \times\,7 \\ \hline \end{array} \qquad \begin{array}{r} 2 \\ \times\,8 \\ \hline \end{array}$$

$$\begin{array}{r} 2 \\ \times\,9 \\ \hline \end{array} \qquad \begin{array}{r} 2 \\ \times\,10 \\ \hline \end{array} \qquad \begin{array}{r} 2 \\ \times\,11 \\ \hline \end{array} \qquad \begin{array}{r} 2 \\ \times\,12 \\ \hline \end{array}$$

Lesson 5

1. Color every 3rd square using a light-colored highlighter, crayon or marker. These numbers will be 3, 6, 9, 12, 15, 18, 21, 24, 27, 30, 33, 36 and 39.

1	2	3	4	5	6	7	8	9	10
11	12	13	14	15	16	17	18	19	20
21	22	23	24	25	26	27	28	29	30
31	32	33	34	35	36	37	38	39	40

2. Read, whisper and shout (really loud) the colored numbers.

Solve these problems:

Use the above number grid to help solve the problems below. For example, to solve 23+10, locate 23 on the number grid and then look down the column. 23+10 =33

$$23+10 = \underline{\hspace{3cm}}$$

$$6+10 = \underline{\hspace{3cm}}$$

$$8+10 = \underline{\hspace{3cm}}$$

$$\begin{array}{r} 1 \\ \times 2 \\ \hline \end{array} \qquad \begin{array}{r} 2 \\ \times 2 \\ \hline \end{array} \qquad \begin{array}{r} 2 \\ \times 6 \\ \hline \end{array} \qquad \begin{array}{r} 4 \\ \times 2 \\ \hline \end{array}$$

$$\begin{array}{r} 2 \\ \times 8 \\ \hline \end{array} \qquad \begin{array}{r} 10 \\ \times 2 \\ \hline \end{array} \qquad \begin{array}{r} 2 \\ \times 3 \\ \hline \end{array} \qquad \begin{array}{r} 5 \\ \times 2 \\ \hline \end{array}$$

$$\begin{array}{r} 2 \\ \times 9 \\ \hline \end{array} \qquad \begin{array}{r} 11 \\ \times 2 \\ \hline \end{array} \qquad \begin{array}{r} 2 \\ \times 12 \\ \hline \end{array} \qquad \begin{array}{r} 7 \\ \times 2 \\ \hline \end{array}$$

Lesson 6

1. Color every 3rd square using a light-colored highlighter, crayon or marker.

1	2	3	4	5	6	7	8	9	10
11	12	13	14	15	16	17	18	19	20
21	22	23	24	25	26	27	28	29	30
31	32	33	34	35	36	37	38	39	40

2. Read, whisper and shout (really loud) the colored numbers.

Solve these problems:

$23+10 =$ _____

$5+10 =$ _____

$28+10 =$ _____

$2+2+2 = \underline{\hspace{2cm}}$ $3\times2 = \underline{\hspace{2cm}}$

$3+3+3 = \underline{\hspace{2cm}}$ $3\times3 = \underline{\hspace{2cm}}$

$4+4+4 = \underline{\hspace{2cm}}$ $3\times4 = \underline{\hspace{2cm}}$

$5+5+5 = \underline{\hspace{2cm}}$ $3\times5 = \underline{\hspace{2cm}}$

$6+6+6 = \underline{\hspace{2cm}}$ $3\times6 = \underline{\hspace{2cm}}$

$7+7+7 = \underline{\hspace{2cm}}$ $3\times7 = \underline{\hspace{2cm}}$

$8+8+8 = \underline{\hspace{2cm}}$ $3\times8 = \underline{\hspace{2cm}}$

$9+9+9 = \underline{\hspace{2cm}}$ $3\times9 = \underline{\hspace{2cm}}$

Lesson 7

1. Color every 3rd square using a light-colored highlighter, crayon or marker.

1	2	3	4	5	6	7	8	9	10
11	12	13	14	15	16	17	18	19	20
21	22	23	24	25	26	27	28	29	30
31	32	33	34	35	36	37	38	39	40

2. Read, whisper and shout (really loud) the colored numbers.

Solve these problems:

$4+10 = $ _____

$16+10 = $ _____

$7+10 = $ _____

$$3 \times 1$$ $$3 \times 2$$ $$3 \times 3$$ $$3 \times 4$$

$$3 \times 5$$ $$3 \times 6$$ $$3 \times 7$$ $$3 \times 8$$

$$3 \times 9$$ $$3 \times 10$$ $$3 \times 11$$ $$3 \times 12$$

Lesson 8

1. Color every 3rd square using a light-colored highlighter, crayon or marker.

1	2	3	4	5	6	7	8	9	10
11	12	13	14	15	16	17	18	19	20
21	22	23	24	25	26	27	28	29	30
31	32	33	34	35	36	37	38	39	40

2. Read, whisper and shout (really loud) the colored numbers.

Solve these problems:

$2+10 = $ _____ $6+10 = $ _____

$8+10 = $ _____ $14+10 = $ _____

$6+10 = $ _____ $7+10 = $ _____

$$\begin{array}{r} 3 \\ \times\,1 \\ \hline \end{array} \qquad \begin{array}{r} 2 \\ \times\,3 \\ \hline \end{array} \qquad \begin{array}{r} 3 \\ \times\,6 \\ \hline \end{array} \qquad \begin{array}{r} 4 \\ \times\,3 \\ \hline \end{array}$$

$$\begin{array}{r} 3 \\ \times\,8 \\ \hline \end{array} \qquad \begin{array}{r} 10 \\ \times\,3 \\ \hline \end{array} \qquad \begin{array}{r} 3 \\ \times\,3 \\ \hline \end{array} \qquad \begin{array}{r} 5 \\ \times\,3 \\ \hline \end{array}$$

$$\begin{array}{r} 3 \\ \times\,9 \\ \hline \end{array} \qquad \begin{array}{r} 11 \\ \times\,3 \\ \hline \end{array} \qquad \begin{array}{r} 3 \\ \times\,12 \\ \hline \end{array} \qquad \begin{array}{r} 7 \\ \times\,3 \\ \hline \end{array}$$

Lesson 9

1. Color every 4th square using a light-colored highlighter, crayon or marker. These numbers will be 4, 8, 12, 16, 20, 24, 28, 32, 36, 40, 44, and 48.

1	2	3	4	5	6	7	8	9	10
11	12	13	14	15	16	17	18	19	20
21	22	23	24	25	26	27	28	29	30
31	32	33	34	35	36	37	38	39	40
41	42	43	44	45	46	47	48	49	50

2. Read, whisper and shout (really loud) the colored numbers.

Solve these problems:

$28+10 =$ ____ $32+10 =$ ____

$31+10 =$ ____ $35+10 =$ ____

$2+10 =$ ____ $15+10 =$ ____

$$\begin{array}{r} 1 \\ \times 2 \\ \hline \end{array} \qquad \begin{array}{r} 3 \\ \times 2 \\ \hline \end{array} \qquad \begin{array}{r} 2 \\ \times 6 \\ \hline \end{array} \qquad \begin{array}{r} 4 \\ \times 3 \\ \hline \end{array}$$

$$\begin{array}{r} 2 \\ \times 8 \\ \hline \end{array} \qquad \begin{array}{r} 10 \\ \times 3 \\ \hline \end{array} \qquad \begin{array}{r} 2 \\ \times 3 \\ \hline \end{array} \qquad \begin{array}{r} 5 \\ \times 3 \\ \hline \end{array}$$

$$\begin{array}{r} 2 \\ \times 9 \\ \hline \end{array} \qquad \begin{array}{r} 11 \\ \times 3 \\ \hline \end{array} \qquad \begin{array}{r} 2 \\ \times 12 \\ \hline \end{array} \qquad \begin{array}{r} 7 \\ \times 3 \\ \hline \end{array}$$

Lesson 10

1. Color every 4th square using a light-colored highlighter, crayon or marker.

1	2	3	4	5	6	7	8	9	10
11	12	13	14	15	16	17	18	19	20
21	22	23	24	25	26	27	28	29	30
31	32	33	34	35	36	37	38	39	40
41	42	43	44	45	46	47	48	49	50

2. Read, whisper and shout (really loud) the colored numbers.

Solve these problems:

$38+10 =$ _____ $30+10 =$ _____

$21+10 =$ _____ $17+10 =$ _____

$31+10 =$ _____ $35+10 =$ _____

$1+1+1+1 =$ _____ $\quad$ $4\times1 =$ _____

$2+2+2+2 =$ _____ $\quad$ $4\times2 =$ _____

$3+3+3+3 =$ _____ $\quad$ $4\times3 =$ _____

$4+4+4+4 =$ _____ $\quad$ $4\times4 =$ _____

$5+5+5+5 =$ _____ $\quad$ $4\times5 =$ _____

$6+6+6+6 =$ _____ $\quad$ $4\times6 =$ _____

$7+7+7+7 =$ _____ $\quad$ $4\times7 =$ _____

$8+8+8+8 =$ _____ $\quad$ $4\times8 =$ _____

$9+9+9+9 =$ _____ $\quad$ $4\times9 =$ _____

$10+10+10+10 =$ _____ $\quad$ $4\times10 =$ _____

$11+11+11+11 =$ _____ $\quad$ $4\times11 =$ _____

Lesson 11

1. Color every 4th square using a light-colored highlighter, crayon or marker.

1	2	3	4	5	6	7	8	9	10
11	12	13	14	15	16	17	18	19	20
21	22	23	24	25	26	27	28	29	30
31	32	33	34	35	36	37	38	39	40
41	42	43	44	45	46	47	48	49	50

2. Read, whisper and shout (really loud) the colored numbers.

Solve these problems:

$23+10 = $ _____

$6+10 = $ _____

$38+10 = $ _____

$$
\begin{array}{r} 4 \\ \times\,1 \\ \hline \end{array}
\qquad
\begin{array}{r} 4 \\ \times\,2 \\ \hline \end{array}
\qquad
\begin{array}{r} 4 \\ \times\,3 \\ \hline \end{array}
\qquad
\begin{array}{r} 4 \\ \times\,4 \\ \hline \end{array}
$$

$$
\begin{array}{r} 4 \\ \times\,5 \\ \hline \end{array}
\qquad
\begin{array}{r} 4 \\ \times\,6 \\ \hline \end{array}
\qquad
\begin{array}{r} 4 \\ \times\,7 \\ \hline \end{array}
\qquad
\begin{array}{r} 4 \\ \times\,8 \\ \hline \end{array}
$$

$$
\begin{array}{r} 4 \\ \times\,9 \\ \hline \end{array}
\qquad
\begin{array}{r} 4 \\ \times\,10 \\ \hline \end{array}
\qquad
\begin{array}{r} 4 \\ \times\,11 \\ \hline \end{array}
\qquad
\begin{array}{r} 4 \\ \times\,12 \\ \hline \end{array}
$$

Lesson 12

1. Color every 4th square using a light-colored highlighter, crayon or marker.

1	2	3	4	5	6	7	8	9	10
11	12	13	14	15	16	17	18	19	20
21	22	23	24	25	26	27	28	29	30
31	32	33	34	35	36	37	38	39	40
41	42	43	44	45	46	47	48	49	50

2. Read, whisper and shout (really loud) the colored numbers.

Solve these problems:

$38+10 = $ _____ $15+10 = $ _____

$26+10 = $ _____ $7+10 = $ _____

$31+10 = $ _____ $4+10 = $ _____

1	2	3	4
$\times 4$	$\times 4$	$\times 4$	$\times 4$

5	6	7	8
$\times 4$	$\times 4$	$\times 4$	$\times 4$

9	10	11	12
$\times 4$	$\times 4$	$\times 4$	$\times 4$

Lesson 13

1. Color every 5th square using a light-colored highlighter, crayon or marker. These numbers will be 5, 10, 15, 20, 25, 30, 35, 40, 45, 50, 55, and 60.

1	2	3	4	5	6	7	8	9	10
11	12	13	14	15	16	17	18	19	20
21	22	23	24	25	26	27	28	29	30
31	32	33	34	35	36	37	38	39	40
41	42	43	44	45	46	47	48	49	50
51	52	53	54	55	56	57	58	59	60

2. Read, whisper and shout (really loud) the colored numbers.

Solve these problems:

$23+10 =$ _____

$56+10 =$ _____

$48+10 =$ _____

$$\begin{array}{r} 4 \\ \times\,1 \\ \hline \end{array} \qquad \begin{array}{r} 4 \\ \times\,2 \\ \hline \end{array} \qquad \begin{array}{r} 4 \\ \times\,3 \\ \hline \end{array} \qquad \begin{array}{r} 4 \\ \times\,4 \\ \hline \end{array}$$

$$\begin{array}{r} 4 \\ \times\,5 \\ \hline \end{array} \qquad \begin{array}{r} 4 \\ \times\,6 \\ \hline \end{array} \qquad \begin{array}{r} 4 \\ \times\,7 \\ \hline \end{array} \qquad \begin{array}{r} 4 \\ \times\,8 \\ \hline \end{array}$$

$$\begin{array}{r} 4 \\ \times\,9 \\ \hline \end{array} \qquad \begin{array}{r} 4 \\ \times\,10 \\ \hline \end{array} \qquad \begin{array}{r} 4 \\ \times\,11 \\ \hline \end{array} \qquad \begin{array}{r} 4 \\ \times\,12 \\ \hline \end{array}$$

Lesson 14

1. Color every 5ᵗʰ square using a light-colored highlighter, crayon or marker.

1	2	3	4	5	6	7	8	9	10
11	12	13	14	15	16	17	18	19	20
21	22	23	24	25	26	27	28	29	30
31	32	33	34	35	36	37	38	39	40
41	42	43	44	45	46	47	48	49	50
51	52	53	54	55	56	57	58	59	60

2. Read, whisper and shout (really loud) the colored numbers.

Solve these problems:

$$11+10 = \underline{\hspace{2cm}}$$

$$35+10 = \underline{\hspace{2cm}}$$

$$49+10 = \underline{\hspace{2cm}}$$

$1+1+1+1+1 = \underline{\hspace{2cm}}$

$5×1 = \underline{\hspace{2cm}}$

$2+2+2+2+2 = \underline{\hspace{2cm}}$

$5×2 = \underline{\hspace{2cm}}$

$3+3+3+3+3 = \underline{\hspace{2cm}}$

$5×3 = \underline{\hspace{2cm}}$

$4+4+4+4+4 = \underline{\hspace{2cm}}$

$5×4 = \underline{\hspace{2cm}}$

$5+5+5+5+5 = \underline{\hspace{2cm}}$

$5×5 = \underline{\hspace{2cm}}$

$6+6+6+6+6 = \underline{\hspace{2cm}}$

$5×6 = \underline{\hspace{2cm}}$

$7+7+7+7+7 = \underline{\hspace{2cm}}$

$5×7 = \underline{\hspace{2cm}}$

$8+8+8+8+8 = \underline{\hspace{2cm}}$

$5×8 = \underline{\hspace{2cm}}$

$9+9+9+9+9 = \underline{\hspace{2cm}}$

$5×9 = \underline{\hspace{2cm}}$

$10+10+10+10+10 = \underline{\hspace{2cm}}$

$5×10 = \underline{\hspace{2cm}}$

$11+11+11+11+11 = \underline{\hspace{2cm}}$

$5×11 = \underline{\hspace{2cm}}$

Lesson 15

1. Color every 5th square using a light-colored highlighter, crayon or marker.

1	2	3	4	5	6	7	8	9	10
11	12	13	14	15	16	17	18	19	20
21	22	23	24	25	26	27	28	29	30
31	32	33	34	35	36	37	38	39	40
41	42	43	44	45	46	47	48	49	50
51	52	53	54	55	56	57	58	59	60

2. Read, whisper and shout (really loud) the colored numbers.

Solve these problems:

$26+10 =$ _____

$16+10 =$ _____

$38+10 =$ _____

$$\begin{array}{r} 5 \\ \times\,1 \\ \hline \end{array} \qquad \begin{array}{r} 5 \\ \times\,2 \\ \hline \end{array} \qquad \begin{array}{r} 5 \\ \times\,3 \\ \hline \end{array} \qquad \begin{array}{r} 5 \\ \times\,4 \\ \hline \end{array}$$

$$\begin{array}{r} 5 \\ \times\,5 \\ \hline \end{array} \qquad \begin{array}{r} 5 \\ \times\,6 \\ \hline \end{array} \qquad \begin{array}{r} 5 \\ \times\,7 \\ \hline \end{array} \qquad \begin{array}{r} 5 \\ \times\,8 \\ \hline \end{array}$$

$$\begin{array}{r} 5 \\ \times\,9 \\ \hline \end{array} \qquad \begin{array}{r} 5 \\ \times\,10 \\ \hline \end{array} \qquad \begin{array}{r} 5 \\ \times\,11 \\ \hline \end{array} \qquad \begin{array}{r} 5 \\ \times\,12 \\ \hline \end{array}$$

Lesson 16

1. Color every 5th square using a light-colored highlighter, crayon or marker.

1	2	3	4	5	6	7	8	9	10
11	12	13	14	15	16	17	18	19	20
21	22	23	24	25	26	27	28	29	30
31	32	33	34	35	36	37	38	39	40
41	42	43	44	45	46	47	48	49	50
51	52	53	54	55	56	57	58	59	60

2. Read, whisper and shout (really loud) the colored numbers.

Solve these problems:

$14+10 =$ _____

$27+10 =$ _____

$50+10 =$ _____

$$\begin{array}{r} 1 \\ \times\,5 \\ \hline \end{array}$$ $$\begin{array}{r} 2 \\ \times\,5 \\ \hline \end{array}$$ $$\begin{array}{r} 3 \\ \times\,5 \\ \hline \end{array}$$ $$\begin{array}{r} 4 \\ \times\,5 \\ \hline \end{array}$$

$$\begin{array}{r} 5 \\ \times\,5 \\ \hline \end{array}$$ $$\begin{array}{r} 6 \\ \times\,5 \\ \hline \end{array}$$ $$\begin{array}{r} 7 \\ \times\,5 \\ \hline \end{array}$$ $$\begin{array}{r} 8 \\ \times\,5 \\ \hline \end{array}$$

$$\begin{array}{r} 9 \\ \times\,5 \\ \hline \end{array}$$ $$\begin{array}{r} 10 \\ \times\,5 \\ \hline \end{array}$$ $$\begin{array}{r} 11 \\ \times\,5 \\ \hline \end{array}$$ $$\begin{array}{r} 12 \\ \times\,5 \\ \hline \end{array}$$

Lesson 17

1. Color every 6th square using a light-colored highlighter, crayon or marker. These numbers will be 6, 12, 18, 24, 30, 36, 42, 48, 54, 60, 66, and 72.

1	2	3	4	5	6	7	8	9	10
11	12	13	14	15	16	17	18	19	20
21	22	23	24	25	26	27	28	29	30
31	32	33	34	35	36	37	38	39	40
41	42	43	44	45	46	47	48	49	50
51	52	53	54	55	56	57	58	59	60
61	62	63	64	65	66	67	68	69	70
71	72	73	74	75	76	77	78	79	80

2. Read, whisper and shout (really loud) the colored numbers.

Solve these problems:

29+10 = _____

26+10 = _____

64+10 = _____

$$\begin{array}{r} 5 \\ \times 1 \\ \hline \end{array} \qquad \begin{array}{r} 5 \\ \times 2 \\ \hline \end{array} \qquad \begin{array}{r} 5 \\ \times 3 \\ \hline \end{array} \qquad \begin{array}{r} 5 \\ \times 4 \\ \hline \end{array}$$

$$\begin{array}{r} 5 \\ \times 5 \\ \hline \end{array} \qquad \begin{array}{r} 5 \\ \times 6 \\ \hline \end{array} \qquad \begin{array}{r} 5 \\ \times 7 \\ \hline \end{array} \qquad \begin{array}{r} 5 \\ \times 8 \\ \hline \end{array}$$

$$\begin{array}{r} 5 \\ \times 9 \\ \hline \end{array} \qquad \begin{array}{r} 5 \\ \times 10 \\ \hline \end{array} \qquad \begin{array}{r} 5 \\ \times 11 \\ \hline \end{array} \qquad \begin{array}{r} 5 \\ \times 12 \\ \hline \end{array}$$

Lesson 18

1. Color every 6th square using a light-colored highlighter, crayon or marker.

1	2	3	4	5	6	7	8	9	10
11	12	13	14	15	16	17	18	19	20
21	22	23	24	25	26	27	28	29	30
31	32	33	34	35	36	37	38	39	40
41	42	43	44	45	46	47	48	49	50
51	52	53	54	55	56	57	58	59	60
61	62	63	64	65	66	67	68	69	70
71	72	73	74	75	76	77	78	79	80

2. Read, whisper and shout (really loud) the colored numbers.

Solve these problems:

$43 + 10 = $ _____

$52 + 10 = $ _____

$38 + 10 = $ _____

$1+1+1+1+1+1 = $ _____

$6×1 = $ _____

$2+2+2+2+2+2 = $ _____

$6×2 = $ _____

$3+3+3+3+3+3 = $ _____

$6×3 = $ _____

$4+4+4+4+4+4 = $ _____

$6×4 = $ _____

$5+5+5+5+5+5 = $ _____

$6×5 = $ _____

$6+6+6+6+6+6 = $ _____

$6×6 = $ _____

$7+7+7+7+7+7 = $ _____

$6×7 = $ _____

$8+8+8+8+8+8 = $ _____

$6×8 = $ _____

$9+9+9+9+9+9 = $ _____

$6×9 = $ _____

$10+10+10+10+10+10 = $ _____

$6×10 = $ ___

$11+11+11+11+11+11 = $ _____

$6×11 = $ ___

Lesson 19

1. Color every 6th square using a light-colored highlighter, crayon or marker.

1	2	3	4	5	6	7	8	9	10
11	12	13	14	15	16	17	18	19	20
21	22	23	24	25	26	27	28	29	30
31	32	33	34	35	36	37	38	39	40
41	42	43	44	45	46	47	48	49	50
51	52	53	54	55	56	57	58	59	60
61	62	63	64	65	66	67	68	69	70
71	72	73	74	75	76	77	78	79	80

2. Read, whisper and shout (really loud) the colored numbers.

Solve these problems:

$33 + 10 = $ _____

$58 + 10 = $ _____

$78 + 10 = $ _____ (Hint: What would be next down the column?)

$$
\begin{array}{r} 6 \\ \times\,1 \\ \hline \end{array}
\qquad
\begin{array}{r} 6 \\ \times\,2 \\ \hline \end{array}
\qquad
\begin{array}{r} 6 \\ \times\,3 \\ \hline \end{array}
\qquad
\begin{array}{r} 6 \\ \times\,4 \\ \hline \end{array}
$$

$$
\begin{array}{r} 6 \\ \times\,5 \\ \hline \end{array}
\qquad
\begin{array}{r} 6 \\ \times\,6 \\ \hline \end{array}
\qquad
\begin{array}{r} 6 \\ \times\,7 \\ \hline \end{array}
\qquad
\begin{array}{r} 6 \\ \times\,8 \\ \hline \end{array}
$$

$$
\begin{array}{r} 6 \\ \times\,9 \\ \hline \end{array}
\qquad
\begin{array}{r} 6 \\ \times\,10 \\ \hline \end{array}
\qquad
\begin{array}{r} 6 \\ \times\,11 \\ \hline \end{array}
\qquad
\begin{array}{r} 6 \\ \times\,12 \\ \hline \end{array}
$$

Lesson 20

1. Color every 6th square using a light-colored highlighter, crayon or marker.

1	2	3	4	5	6	7	8	9	10
11	12	13	14	15	16	17	18	19	20
21	22	23	24	25	26	27	28	29	30
31	32	33	34	35	36	37	38	39	40
41	42	43	44	45	46	47	48	49	50
51	52	53	54	55	56	57	58	59	60
61	62	63	64	65	66	67	68	69	70
71	72	73	74	75	76	77	78	79	80

2. Read, whisper and shout (really loud) the colored numbers.

Solve these problems:

$23 + 10 =$ _____

$56 + 10 =$ _____

$68 + 10 =$ _____

$$
\begin{array}{cccc}
1 & 2 & 3 & 4 \\
\times\,6 & \times\,6 & \times\,6 & \times\,6 \\
\hline
\end{array}
$$

$$
\begin{array}{cccc}
5 & 6 & 7 & 8 \\
\times\,6 & \times\,6 & \times\,6 & \times\,6 \\
\hline
\end{array}
$$

$$
\begin{array}{cccc}
9 & 10 & 11 & 12 \\
\times\,6 & \times\,6 & \times\,6 & \times\,6 \\
\hline
\end{array}
$$

Lesson 21

1. Color every 7th square using a light-colored highlighter, crayon or marker. These numbers will be 7, 14, 21, 28, 35, 42, 49, 56, 63, 70, 77, and 84.

1	2	3	4	5	6	7	8	9	10
11	12	13	14	15	16	17	18	19	20
21	22	23	24	25	26	27	28	29	30
31	32	33	34	35	36	37	38	39	40
41	42	43	44	45	46	47	48	49	50
51	52	53	54	55	56	57	58	59	60
61	62	63	64	65	66	67	68	69	70
71	72	73	74	75	76	77	78	79	80
81	82	83	84	85	86	87	88	89	90
91	92	93	94	95	96	97	98	99	100

2. Read, whisper and shout (really loud) the colored numbers.

Solve these problems:

24+10+10 = _____   48+10 = _____

58+10+10 = _____   49+10 = _____

$$\begin{array}{r} 6 \\ \times\,1 \\ \hline \end{array} \qquad \begin{array}{r} 6 \\ \times\,2 \\ \hline \end{array} \qquad \begin{array}{r} 6 \\ \times\,3 \\ \hline \end{array} \qquad \begin{array}{r} 6 \\ \times\,4 \\ \hline \end{array}$$

$$\begin{array}{r} 6 \\ \times\,5 \\ \hline \end{array} \qquad \begin{array}{r} 6 \\ \times\,6 \\ \hline \end{array} \qquad \begin{array}{r} 6 \\ \times\,7 \\ \hline \end{array} \qquad \begin{array}{r} 6 \\ \times\,8 \\ \hline \end{array}$$

$$\begin{array}{r} 6 \\ \times\,9 \\ \hline \end{array} \qquad \begin{array}{r} 6 \\ \times\,10 \\ \hline \end{array} \qquad \begin{array}{r} 6 \\ \times\,11 \\ \hline \end{array} \qquad \begin{array}{r} 6 \\ \times\,12 \\ \hline \end{array}$$

Lesson 22

1. Color every 7th square using a light-colored highlighter, crayon or marker.

1	2	3	4	5	6	7	8	9	10
11	12	13	14	15	16	17	18	19	20
21	22	23	24	25	26	27	28	29	30
31	32	33	34	35	36	37	38	39	40
41	42	43	44	45	46	47	48	49	50
51	52	53	54	55	56	57	58	59	60
61	62	63	64	65	66	67	68	69	70
71	72	73	74	75	76	77	78	79	80
81	82	83	84	85	86	87	88	89	90
91	92	93	94	95	96	97	98	99	100

2. Read, whisper and shout (really loud) the colored numbers.

Solve these problems:

$12+10+10 =$ _____ $87+10 =$ _____

$45+10+10 =$ _____ $65+10 =$ _____

$1+1+1+1+1+1+1 = \underline{\hphantom{000}}$ $\qquad$ $7\times1 = \underline{\hphantom{000}}$

$2+2+2+2+2+2+2 = \underline{\hphantom{000}}$ $\qquad$ $7\times2 = \underline{\hphantom{000}}$

$3+3+3+3+3+3+3 = \underline{\hphantom{000}}$ $\qquad$ $7\times3 = \underline{\hphantom{000}}$

$4+4+4+4+4+4+4 = \underline{\hphantom{000}}$ $\qquad$ $7\times4 = \underline{\hphantom{000}}$

$5+5+5+5+5+5+5 = \underline{\hphantom{000}}$ $\qquad$ $7\times5 = \underline{\hphantom{000}}$

$6+6+6+6+6+6+6 = \underline{\hphantom{000}}$ $\qquad$ $7\times6 = \underline{\hphantom{000}}$

$7+7+7+7+7+7+7 = \underline{\hphantom{000}}$ $\qquad$ $7\times7 = \underline{\hphantom{000}}$

$8+8+8+8+8+8+8 = \underline{\hphantom{000}}$ $\qquad$ $7\times8 = \underline{\hphantom{000}}$

$9+9+9+9+9+9+9 = \underline{\hphantom{000}}$ $\qquad$ $7\times9 = \underline{\hphantom{000}}$

$10+10+10+10+10+10+10 = \underline{\hphantom{000}}$ $\quad$ $7\times10 = \underline{\hphantom{000}}$

$11+11+11+11+11+11+11 = \underline{\hphantom{000}}$ $\quad$ $7\times11 = \underline{\hphantom{000}}$

Lesson 23

1. Color every 7th square using a light-colored highlighter, crayon or marker.

1	2	3	4	5	6	7	8	9	10
11	12	13	14	15	16	17	18	19	20
21	22	23	24	25	26	27	28	29	30
31	32	33	34	35	36	37	38	39	40
41	42	43	44	45	46	47	48	49	50
51	52	53	54	55	56	57	58	59	60
61	62	63	64	65	66	67	68	69	70
71	72	73	74	75	76	77	78	79	80
81	82	83	84	85	86	87	88	89	90
91	92	93	94	95	96	97	98	99	100

2. Read, whisper and shout (really loud) the colored numbers.

Solve these problems:

$64+10+10 =$ _____ $88+10 =$ _____

$57+10+10 =$ _____ $45+10 =$ _____

$$\begin{array}{r} 7 \\ \times\,1 \\ \hline \end{array} \qquad \begin{array}{r} 7 \\ \times\,2 \\ \hline \end{array} \qquad \begin{array}{r} 7 \\ \times\,3 \\ \hline \end{array} \qquad \begin{array}{r} 7 \\ \times\,4 \\ \hline \end{array}$$

$$\begin{array}{r} 7 \\ \times\,5 \\ \hline \end{array} \qquad \begin{array}{r} 7 \\ \times\,6 \\ \hline \end{array} \qquad \begin{array}{r} 7 \\ \times\,7 \\ \hline \end{array} \qquad \begin{array}{r} 7 \\ \times\,8 \\ \hline \end{array}$$

$$\begin{array}{r} 7 \\ \times\,9 \\ \hline \end{array} \qquad \begin{array}{r} 7 \\ \times\,10 \\ \hline \end{array} \qquad \begin{array}{r} 7 \\ \times\,11 \\ \hline \end{array} \qquad \begin{array}{r} 7 \\ \times\,12 \\ \hline \end{array}$$

Lesson 24

1. Color every 7th square using a light-colored highlighter, crayon or marker.

1	2	3	4	5	6	7	8	9	10
11	12	13	14	15	16	17	18	19	20
21	22	23	24	25	26	27	28	29	30
31	32	33	34	35	36	37	38	39	40
41	42	43	44	45	46	47	48	49	50
51	52	53	54	55	56	57	58	59	60
61	62	63	64	65	66	67	68	69	70
71	72	73	74	75	76	77	78	79	80
81	82	83	84	85	86	87	88	89	90
91	92	93	94	95	96	97	98	99	100

2. Read, whisper and shout (really loud) the colored numbers.

Solve these problems:

$74+10+10 =$ _____ $86+10 =$ _____

$66+10+10 =$ _____ $82+10 =$ _____

1	2	3	4
× 7	× 7	× 7	× 7

5	6	7	8
× 7	× 7	× 7	× 7

9	10	11	12
× 7	× 7	× 7	× 7

Lesson 25

1. Color every 8th square using a light-colored highlighter, crayon or marker. These numbers will be 8, 16, 24, 32, 40, 48, 56, 64, 72, 80, 88, and 96.

1	2	3	4	5	6	7	8	9	10
11	12	13	14	15	16	17	18	19	20
21	22	23	24	25	26	27	28	29	30
31	32	33	34	35	36	37	38	39	40
41	42	43	44	45	46	47	48	49	50
51	52	53	54	55	56	57	58	59	60
61	62	63	64	65	66	67	68	69	70
71	72	73	74	75	76	77	78	79	80
81	82	83	84	85	86	87	88	89	90
91	92	93	94	95	96	97	98	99	100

2. Read, whisper and shout (really loud) the colored numbers.

Solve these problems:

$61+10+10 = $ _____ 　　　　$63+10 = $ _____

$58+10+10 = $ _____ 　　　　$40+10 = $ _____

$$\begin{array}{r} 7 \\ \times\,1 \\ \hline \end{array} \qquad \begin{array}{r} 7 \\ \times\,2 \\ \hline \end{array} \qquad \begin{array}{r} 7 \\ \times\,3 \\ \hline \end{array} \qquad \begin{array}{r} 7 \\ \times\,4 \\ \hline \end{array}$$

$$\begin{array}{r} 7 \\ \times\,5 \\ \hline \end{array} \qquad \begin{array}{r} 7 \\ \times\,6 \\ \hline \end{array} \qquad \begin{array}{r} 7 \\ \times\,7 \\ \hline \end{array} \qquad \begin{array}{r} 7 \\ \times\,8 \\ \hline \end{array}$$

$$\begin{array}{r} 7 \\ \times\,9 \\ \hline \end{array} \qquad \begin{array}{r} 7 \\ \times\,10 \\ \hline \end{array} \qquad \begin{array}{r} 7 \\ \times\,11 \\ \hline \end{array} \qquad \begin{array}{r} 7 \\ \times\,12 \\ \hline \end{array}$$

Lesson 26

1. Color every 8th square using a light-colored highlighter, crayon or marker.

1	2	3	4	5	6	7	8	9	10
11	12	13	14	15	16	17	18	19	20
21	22	23	24	25	26	27	28	29	30
31	32	33	34	35	36	37	38	39	40
41	42	43	44	45	46	47	48	49	50
51	52	53	54	55	56	57	58	59	60
61	62	63	64	65	66	67	68	69	70
71	72	73	74	75	76	77	78	79	80
81	82	83	84	85	86	87	88	89	90
91	92	93	94	95	96	97	98	99	100

2. Read, whisper and shout (really loud) the colored numbers.

Solve these problems:

$33+10+10 =$ _____ $77+10 =$ _____

$13+10+10 =$ _____ $37+10 =$ _____

$1+1+1+1+1+1+1+1 = \underline{}$

$8 \times 1 = \underline{}$

$2+2+2+2+2+2+2+2 = \underline{}$

$8 \times 2 = \underline{}$

$3+3+3+3+3+3+3+3 = \underline{}$

$8 \times 3 = \underline{}$

$4+4+4+4+4+4+4+4 = \underline{}$

$8 \times 4 = \underline{}$

$5+5+5+5+5+5+5+5 = \underline{}$

$8 \times 5 = \underline{}$

$6+6+6+6+6+6+6+6 = \underline{}$

$8 \times 6 = \underline{}$

$7+7+7+7+7+7+7+7 = \underline{}$

$8 \times 7 = \underline{}$

$8+8+8+8+8+8+8+8 = \underline{}$

$8 \times 8 = \underline{}$

$9+9+9+9+9+9+9+9 = \underline{}$

$8 \times 9 = \underline{}$

$10+10+10+10+10+10+10+10 = \underline{}$

$8 \times 10 = \underline{}$

$11+11+11+11+11+11+11+11 = \underline{}$

$8 \times 11 = \underline{}$

Lesson 27

1. Color every 8th square using a light-colored highlighter, crayon or marker.

1	2	3	4	5	6	7	8	9	10
11	12	13	14	15	16	17	18	19	20
21	22	23	24	25	26	27	28	29	30
31	32	33	34	35	36	37	38	39	40
41	42	43	44	45	46	47	48	49	50
51	52	53	54	55	56	57	58	59	60
61	62	63	64	65	66	67	68	69	70
71	72	73	74	75	76	77	78	79	80
81	82	83	84	85	86	87	88	89	90
91	92	93	94	95	96	97	98	99	100

2. Read, whisper and shout (really loud) the colored numbers.

Solve these problems:

$3+10+10+10 =$ _____ $24+10 =$ _____

$7+10+10+10 =$ _____ $18+10 =$ _____

$$\begin{array}{r} 8 \\ \times\,1 \\ \hline \end{array} \qquad \begin{array}{r} 8 \\ \times\,2 \\ \hline \end{array} \qquad \begin{array}{r} 8 \\ \times\,3 \\ \hline \end{array} \qquad \begin{array}{r} 8 \\ \times\,4 \\ \hline \end{array}$$

$$\begin{array}{r} 8 \\ \times\,5 \\ \hline \end{array} \qquad \begin{array}{r} 8 \\ \times\,6 \\ \hline \end{array} \qquad \begin{array}{r} 8 \\ \times\,7 \\ \hline \end{array} \qquad \begin{array}{r} 8 \\ \times\,8 \\ \hline \end{array}$$

$$\begin{array}{r} 8 \\ \times\,9 \\ \hline \end{array} \qquad \begin{array}{r} 8 \\ \times\,10 \\ \hline \end{array} \qquad \begin{array}{r} 8 \\ \times\,11 \\ \hline \end{array} \qquad \begin{array}{r} 8 \\ \times\,12 \\ \hline \end{array}$$

Lesson 28

1. Color every 8th square using a light-colored highlighter, crayon or marker.

1	2	3	4	5	6	7	8	9	10
11	12	13	14	15	16	17	18	19	20
21	22	23	24	25	26	27	28	29	30
31	32	33	34	35	36	37	38	39	40
41	42	43	44	45	46	47	48	49	50
51	52	53	54	55	56	57	58	59	60
61	62	63	64	65	66	67	68	69	70
71	72	73	74	75	76	77	78	79	80
81	82	83	84	85	86	87	88	89	90
91	92	93	94	95	96	97	98	99	100

2. Read, whisper and shout (really loud) the colored numbers.

Solve these problems:

$14+10+10+10 = $ _____ $95+10 = $ _____

$13+10+10+10 = $ _____ $77+10 = $ _____

$$\begin{array}{r} 1 \\ \times\ 8 \\ \hline \end{array}$$
$$\begin{array}{r} 2 \\ \times\ 8 \\ \hline \end{array}$$
$$\begin{array}{r} 3 \\ \times\ 8 \\ \hline \end{array}$$
$$\begin{array}{r} 4 \\ \times\ 8 \\ \hline \end{array}$$

$$\begin{array}{r} 5 \\ \times\ 8 \\ \hline \end{array}$$
$$\begin{array}{r} 6 \\ \times\ 8 \\ \hline \end{array}$$
$$\begin{array}{r} 7 \\ \times\ 8 \\ \hline \end{array}$$
$$\begin{array}{r} 8 \\ \times\ 8 \\ \hline \end{array}$$

$$\begin{array}{r} 9 \\ \times\ 8 \\ \hline \end{array}$$
$$\begin{array}{r} 10 \\ \times\ 8 \\ \hline \end{array}$$
$$\begin{array}{r} 11 \\ \times\ 8 \\ \hline \end{array}$$
$$\begin{array}{r} 12 \\ \times\ 8 \\ \hline \end{array}$$

Lesson 29

1. Color every 9th square using a light-colored highlighter, crayon or marker. These numbers will be 9, 18, 27, 36, 45, 54, 63, 72, 81, 90, 99, and 108.

1	2	3	4	5	6	7	8	9	10
11	12	13	14	15	16	17	18	19	20
21	22	23	24	25	26	27	28	29	30
31	32	33	34	35	36	37	38	39	40
41	42	43	44	45	46	47	48	49	50
51	52	53	54	55	56	57	58	59	60
61	62	63	64	65	66	67	68	69	70
71	72	73	74	75	76	77	78	79	80
81	82	83	84	85	86	87	88	89	90
91	92	93	94	95	96	97	98	99	100

2. Read, whisper and shout (really loud) the colored numbers.

Solve these problems:

$61+10+10+10 =$ _____ $33+10 =$ _____

$59+10+10+10 =$ _____ $44+10 =$ _____

$$8 \times 1 \qquad 8 \times 2 \qquad 8 \times 3 \qquad 8 \times 4$$

$$8 \times 5 \qquad 8 \times 6 \qquad 8 \times 7 \qquad 8 \times 8$$

$$8 \times 9 \qquad 8 \times 10 \qquad 8 \times 11 \qquad 8 \times 12$$

Lesson 30

1. Color every 9ᵗʰ square using a light-colored highlighter, crayon or marker.

1	2	3	4	5	6	7	8	9	10
11	12	13	14	15	16	17	18	19	20
21	22	23	24	25	26	27	28	29	30
31	32	33	34	35	36	37	38	39	40
41	42	43	44	45	46	47	48	49	50
51	52	53	54	55	56	57	58	59	60
61	62	63	64	65	66	67	68	69	70
71	72	73	74	75	76	77	78	79	80
81	82	83	84	85	86	87	88	89	90
91	92	93	94	95	96	97	98	99	100

2. Read, whisper and shout (really loud) the colored numbers.

Solve these problems:

$12+10+10+10 = \underline{\hspace{1cm}}$ $56+10 = \underline{\hspace{1cm}}$

$64+10+10+10 = \underline{\hspace{1cm}}$ $77+10 = \underline{\hspace{1cm}}$

$1+1+1+1+1+1+1+1+1 =$ ___ $9\times1 =$ ___

$2+2+2+2+2+2+2+2+2 =$ ___ $9\times2 =$ ___

$3+3+3+3+3+3+3+3+3 =$ ___ $9\times3 =$ ___

$4+4+4+4+4+4+4+4+4 =$ ___ $9\times4 =$ ___

$5+5+5+5+5+5+5+5+5 =$ ___ $9\times5 =$ ___

$6+6+6+6+6+6+6+6+6 =$ ___ $9\times6 =$ ___

$7+7+7+7+7+7+7+7+7 =$ ___ $9\times7 =$ ___

$8+8+8+8+8+8+8+8+8 =$ ___ $9\times8 =$ ___

$9+9+9+9+9+9+9+9+9 =$ ___ $9\times9 =$ ___

$10+10+10+10+10+10+10+10+10 =$ ___
$9\times10 =$ ___
$11+11+11+11+11+11+11+11+11 =$ ___
$9\times11 =$ ___

Lesson 31

1. Color every 9th square using a light-colored highlighter, crayon or marker.

1	2	3	4	5	6	7	8	9	10
11	12	13	14	15	16	17	18	19	20
21	22	23	24	25	26	27	28	29	30
31	32	33	34	35	36	37	38	39	40
41	42	43	44	45	46	47	48	49	50
51	52	53	54	55	56	57	58	59	60
61	62	63	64	65	66	67	68	69	70
71	72	73	74	75	76	77	78	79	80
81	82	83	84	85	86	87	88	89	90
91	92	93	94	95	96	97	98	99	100

2. Read, whisper and shout (really loud) the colored numbers.

Solve these problems:

$11+10+10+10 =$ _____ $86+10 =$ _____

$47+10+10+10 =$ _____ $55+10 =$ _____

$$
\begin{array}{r} 9 \\ \times\,1 \\ \hline \end{array}
\qquad
\begin{array}{r} 9 \\ \times\,2 \\ \hline \end{array}
\qquad
\begin{array}{r} 9 \\ \times\,3 \\ \hline \end{array}
\qquad
\begin{array}{r} 9 \\ \times\,4 \\ \hline \end{array}
$$

$$
\begin{array}{r} 9 \\ \times\,5 \\ \hline \end{array}
\qquad
\begin{array}{r} 9 \\ \times\,6 \\ \hline \end{array}
\qquad
\begin{array}{r} 9 \\ \times\,7 \\ \hline \end{array}
\qquad
\begin{array}{r} 9 \\ \times\,8 \\ \hline \end{array}
$$

$$
\begin{array}{r} 9 \\ \times\,9 \\ \hline \end{array}
\qquad
\begin{array}{r} 9 \\ \times\,10 \\ \hline \end{array}
\qquad
\begin{array}{r} 9 \\ \times\,11 \\ \hline \end{array}
\qquad
\begin{array}{r} 9 \\ \times\,12 \\ \hline \end{array}
$$

Lesson 32

1. Color every 9th square using a light-colored highlighter, crayon or marker.

1	2	3	4	5	6	7	8	9	10
11	12	13	14	15	16	17	18	19	20
21	22	23	24	25	26	27	28	29	30
31	32	33	34	35	36	37	38	39	40
41	42	43	44	45	46	47	48	49	50
51	52	53	54	55	56	57	58	59	60
61	62	63	64	65	66	67	68	69	70
71	72	73	74	75	76	77	78	79	80
81	82	83	84	85	86	87	88	89	90
91	92	93	94	95	96	97	98	99	100

2. Read, whisper and shout (really loud) the colored numbers.

Solve these problems:

$45+10+10+10 =$ _____ $92+10 =$ _____

$37+10+10+10 =$ _____ $21+10 =$ _____

1	2	3	4
× 9	× 9	× 9	× 9
5	6	7	8
× 9	× 9	× 9	× 9
9	10	11	12
× 9	× 9	× 9	× 9

Lesson 33

Multiplication by Zero and One

Zero multiplied by any number is zero. One multiplied by any number is that number.

For example:

$$\begin{array}{r} 5 \\ \times\,0 \\ \hline 0 \end{array} \qquad \begin{array}{r} 5 \\ \times\,1 \\ \hline 5 \end{array} \qquad \begin{array}{r} 3 \\ \times\,0 \\ \hline 0 \end{array} \qquad \begin{array}{r} 1 \\ \times\,3 \\ \hline 3 \end{array} \qquad \begin{array}{r} 0 \\ \times\,6 \\ \hline 0 \end{array}$$

Now try these problems:

$$\begin{array}{r} 5 \\ \times\,0 \\ \hline \end{array} \qquad \begin{array}{r} 5 \\ \times\,1 \\ \hline \end{array} \qquad \begin{array}{r} 8 \\ \times\,0 \\ \hline \end{array} \qquad \begin{array}{r} 9 \\ \times\,1 \\ \hline \end{array} \qquad \begin{array}{r} 3 \\ \times\,0 \\ \hline \end{array}$$

$5+5+5 \quad = \underline{\quad} \qquad 3\times5 = \underline{\quad} \quad$ (three fives)

$5+5 \quad = \underline{\quad} \qquad 2\times5 = \underline{\quad} \quad$ (two fives)

$5 \quad = \underline{\quad} \qquad 1\times5 = \underline{\quad} \quad$ (one five)

no fives $\quad = \underline{\quad} \qquad 0\times5 = \underline{\quad} \quad$ (no fives)

$$\begin{array}{r} 5 \\ \times\,0 \\ \hline \end{array} \qquad \begin{array}{r} 1 \\ \times\,7 \\ \hline \end{array} \qquad \begin{array}{r} 8 \\ \times\,0 \\ \hline \end{array} \qquad \begin{array}{r} 1 \\ \times\,8 \\ \hline \end{array} \qquad \begin{array}{r} 0 \\ \times\,0 \\ \hline \end{array}$$

$$\begin{array}{r} 1 \\ \times\,3 \\ \hline \end{array} \qquad \begin{array}{r} 7 \\ \times\,0 \\ \hline \end{array} \qquad \begin{array}{r} 4 \\ \times\,1 \\ \hline \end{array} \qquad \begin{array}{r} 0 \\ \times\,6 \\ \hline \end{array} \qquad \begin{array}{r} 5 \\ \times\,1 \\ \hline \end{array}$$

$$\begin{array}{r} 5 \\ \times\,0 \\ \hline \end{array} \qquad \begin{array}{r} 3 \\ \times\,1 \\ \hline \end{array} \qquad \begin{array}{r} 0 \\ \times\,8 \\ \hline \end{array} \qquad \begin{array}{r} 1 \\ \times\,2 \\ \hline \end{array} \qquad \begin{array}{r} 5 \\ \times\,1 \\ \hline \end{array}$$

$$\begin{array}{r} 1 \\ \times\,5 \\ \hline \end{array} \qquad \begin{array}{r} 6 \\ \times\,0 \\ \hline \end{array} \qquad \begin{array}{r} 3 \\ \times\,1 \\ \hline \end{array} \qquad \begin{array}{r} 4 \\ \times\,1 \\ \hline \end{array} \qquad \begin{array}{r} 0 \\ \times\,3 \\ \hline \end{array}$$

$$\begin{array}{r} 2 \\ \times\,0 \\ \hline \end{array} \qquad \begin{array}{r} 8 \\ \times\,1 \\ \hline \end{array} \qquad \begin{array}{r} 0 \\ \times\,5 \\ \hline \end{array} \qquad \begin{array}{r} 7 \\ \times\,1 \\ \hline \end{array} \qquad \begin{array}{r} 9 \\ \times\,0 \\ \hline \end{array}$$

Lesson 34

1. Color every 10th square using a light-colored highlighter, crayon or marker. These numbers will be 10, 20, 30, 40, 50, 60, 70, 80, 90, 100, 110, and 120.

1	2	3	4	5	6	7	8	9	10
11	12	13	14	15	16	17	18	19	20
21	22	23	24	25	26	27	28	29	30
31	32	33	34	35	36	37	38	39	40
41	42	43	44	45	46	47	48	49	50
51	52	53	54	55	56	57	58	59	60
61	62	63	64	65	66	67	68	69	70
71	72	73	74	75	76	77	78	79	80
81	82	83	84	85	86	87	88	89	90
91	92	93	94	95	96	97	98	99	100

2. Read, whisper and shout (really loud) the colored numbers.

Solve these problems:

$22+10+10+10 =$ _____ $63+10 =$ _____

$44+10+10+10 =$ _____ $59+10 =$ _____

$$\begin{array}{r} 9 \\ \times\,1 \\ \hline \end{array} \qquad \begin{array}{r} 9 \\ \times\,2 \\ \hline \end{array} \qquad \begin{array}{r} 9 \\ \times\,3 \\ \hline \end{array} \qquad \begin{array}{r} 9 \\ \times\,4 \\ \hline \end{array}$$

$$\begin{array}{r} 9 \\ \times\,5 \\ \hline \end{array} \qquad \begin{array}{r} 9 \\ \times\,6 \\ \hline \end{array} \qquad \begin{array}{r} 9 \\ \times\,7 \\ \hline \end{array} \qquad \begin{array}{r} 9 \\ \times\,8 \\ \hline \end{array}$$

$$\begin{array}{r} 9 \\ \times\,9 \\ \hline \end{array} \qquad \begin{array}{r} 9 \\ \times\,10 \\ \hline \end{array} \qquad \begin{array}{r} 9 \\ \times\,11 \\ \hline \end{array} \qquad \begin{array}{r} 9 \\ \times\,12 \\ \hline \end{array}$$

Lesson 35

1. Color every 10th using a light-colored highlighter, crayon or marker.

1	2	3	4	5	6	7	8	9	10
11	12	13	14	15	16	17	18	19	20
21	22	23	24	25	26	27	28	29	30
31	32	33	34	35	36	37	38	39	40
41	42	43	44	45	46	47	48	49	50
51	52	53	54	55	56	57	58	59	60
61	62	63	64	65	66	67	68	69	70
71	72	73	74	75	76	77	78	79	80
81	82	83	84	85	86	87	88	89	90
91	92	93	94	95	96	97	98	99	100

2. Read, whisper and shout (really loud) the colored numbers.

Solve these problems:

$12+10+10+10 = $ _____ $74+10 = $ _____

$64+10+10+10 = $ _____ $98+10 = $ _____

$1+1+1+1+1+1+1+1+1+1=$___ $10\times1=$

$2+2+2+2+2+2+2+2+2+2=$___ $10\times2=$

$3+3+3+3+3+3+3+3+3+3=$___ $10\times3=$

$4+4+4+4+4+4+4+4+4+4=$___ $10\times4=$

$5+5+5+5+5+5+5+5+5+5=$___ $10\times5=$

$6+6+6+6+6+6+6+6+6+6=$___ $10\times6=$

$7+7+7+7+7+7+7+7+7+7=$___ $10\times7=$

$8+8+8+8+8+8+8+8+8+8=$___ $10\times8=$

$9+9+9+9+9+9+9+9+9+9=$___ $10\times9=$

$10+10+10+10+10+10+10+10+10+10=$___
$10\times10=$ ___
$11+11+11+11+11+11+11+11+11+11=$___
$10\times11=$ ___

Lesson 36

1. Color every 10th square using a light-colored highlighter, crayon or marker.

1	2	3	4	5	6	7	8	9	10
11	12	13	14	15	16	17	18	19	20
21	22	23	24	25	26	27	28	29	30
31	32	33	34	35	36	37	38	39	40
41	42	43	44	45	46	47	48	49	50
51	52	53	54	55	56	57	58	59	60
61	62	63	64	65	66	67	68	69	70
71	72	73	74	75	76	77	78	79	80
81	82	83	84	85	86	87	88	89	90
91	92	93	94	95	96	97	98	99	100

2. Read, whisper and shout (really loud) the colored numbers.

Solve these problems:

$11+10+10+10 = $ _____ $7+10 = $ _____

$17+10+10+10 = $ _____ $2+10 = $ _____

$$\begin{array}{r} 10 \\ \times\,1 \\ \hline \end{array} \qquad \begin{array}{r} 10 \\ \times\,2 \\ \hline \end{array} \qquad \begin{array}{r} 10 \\ \times\,3 \\ \hline \end{array} \qquad \begin{array}{r} 10 \\ \times\,4 \\ \hline \end{array}$$

$$\begin{array}{r} 10 \\ \times\,5 \\ \hline \end{array} \qquad \begin{array}{r} 10 \\ \times\,6 \\ \hline \end{array} \qquad \begin{array}{r} 10 \\ \times\,7 \\ \hline \end{array} \qquad \begin{array}{r} 10 \\ \times\,8 \\ \hline \end{array}$$

$$\begin{array}{r} 10 \\ \times\,9 \\ \hline \end{array} \qquad \begin{array}{r} 10 \\ \times\,10 \\ \hline \end{array} \qquad \begin{array}{r} 10 \\ \times\,11 \\ \hline \end{array} \qquad \begin{array}{r} 10 \\ \times\,12 \\ \hline \end{array}$$

Lesson 37

1. Color every 10th square using a light-colored highlighter, crayon or marker.

1	2	3	4	5	6	7	8	9	10
11	12	13	14	15	16	17	18	19	20
21	22	23	24	25	26	27	28	29	30
31	32	33	34	35	36	37	38	39	40
41	42	43	44	45	46	47	48	49	50
51	52	53	54	55	56	57	58	59	60
61	62	63	64	65	66	67	68	69	70
71	72	73	74	75	76	77	78	79	80
81	82	83	84	85	86	87	88	89	90
91	92	93	94	95	96	97	98	99	100

2. Read, whisper and shout (really loud) the colored numbers.

Solve these problems:

$4+10+10+10 = $ ____ $10+10 = $ ____

$17+10+10+10 = $ ____ $22+10 = $ ____

$$\begin{array}{r} 1 \\ \times\ 10 \\ \hline \end{array} \qquad \begin{array}{r} 2 \\ \times\ 10 \\ \hline \end{array} \qquad \begin{array}{r} 3 \\ \times\ 10 \\ \hline \end{array} \qquad \begin{array}{r} 4 \\ \times\ 10 \\ \hline \end{array}$$

$$\begin{array}{r} 5 \\ \times\ 10 \\ \hline \end{array} \qquad \begin{array}{r} 6 \\ \times\ 10 \\ \hline \end{array} \qquad \begin{array}{r} 7 \\ \times\ 10 \\ \hline \end{array} \qquad \begin{array}{r} 8 \\ \times\ 10 \\ \hline \end{array}$$

$$\begin{array}{r} 9 \\ \times 10 \\ \hline \end{array} \qquad \begin{array}{r} 10 \\ \times 10 \\ \hline \end{array} \qquad \begin{array}{r} 11 \\ \times 10 \\ \hline \end{array} \qquad \begin{array}{r} 12 \\ \times 10 \\ \hline \end{array}$$

Lesson 38

1. Color every 11th square using a light-colored highlighter, crayon or marker. These numbers will be 11, 22, 33, 44, 55, 66, 77, 88, 99, 110, 121, and 132.

1	2	3	4	5	6	7	8	9	10
11	12	13	14	15	16	17	18	19	20
21	22	23	24	25	26	27	28	29	30
31	32	33	34	35	36	37	38	39	40
41	42	43	44	45	46	47	48	49	50
51	52	53	54	55	56	57	58	59	60
61	62	63	64	65	66	67	68	69	70
71	72	73	74	75	76	77	78	79	80
81	82	83	84	85	86	87	88	89	90
91	92	93	94	95	96	97	98	99	100

2. Read, whisper and shout (really loud) the colored numbers.

Solve these problems:

$45+10+10+10 =$ _____ $83+10 =$ _____

$3+10+10+10 =$ _____ $80+10 =$ _____

$$
\begin{array}{cccc}
10 & 10 & 10 & 10 \\
\underline{\times 1} & \underline{\times 2} & \underline{\times 3} & \underline{\times 4}
\end{array}
$$

$$
\begin{array}{cccc}
10 & 10 & 10 & 10 \\
\underline{\times 5} & \underline{\times 6} & \underline{\times 7} & \underline{\times 8}
\end{array}
$$

$$
\begin{array}{cccc}
10 & 10 & 10 & 10 \\
\underline{\times 9} & \underline{\times 10} & \underline{\times 11} & \underline{\times 12}
\end{array}
$$

Lesson 39

1. Color every 11th square using a light-colored highlighter, crayon or marker.

1	2	3	4	5	6	7	8	9	10
11	12	13	14	15	16	17	18	19	20
21	22	23	24	25	26	27	28	29	30
31	32	33	34	35	36	37	38	39	40
41	42	43	44	45	46	47	48	49	50
51	52	53	54	55	56	57	58	59	60
61	62	63	64	65	66	67	68	69	70
71	72	73	74	75	76	77	78	79	80
81	82	83	84	85	86	87	88	89	90
91	92	93	94	95	96	97	98	99	100

2. Read, whisper and shout (really loud) the colored numbers.

Solve these problems:

$27+10+10+10 = $ _____ $74+10 = $ _____

$15+10+10+10 = $ _____ $89+10 = $ _____

$1+1+1+1+1+1+1+1+1+1+1 = \underline{\quad}$ $11 \times 1 =$

$2+2+2+2+2+2+2+2+2+2+2 = \underline{\quad}$ $11 \times 2 =$

$3+3+3+3+3+3+3+3+3+3+3 = \underline{\quad}$ $11 \times 3 =$

$4+4+4+4+4+4+4+4+4+4+4 = \underline{\quad}$ $11 \times 4 =$

$5+5+5+5+5+5+5+5+5+5+5 = \underline{\quad}$ $11 \times 5 =$

$6+6+6+6+6+6+6+6+6+6+6 = \underline{\quad}$ $11 \times 6 =$

$7+7+7+7+7+7+7+7+7+7+7 = \underline{\quad}$ $11 \times 7 =$

$8+8+8+8+8+8+8+8+8+8+8 = \underline{\quad}$ $11 \times 8 =$

$9+9+9+9+9+9+9+9+9+9+9 = \underline{\quad}$ $11 \times 9 =$

$10+10+10+10+10+10+10+10+10+10+10$
$= \underline{\quad}$ $11 \times 10 = \underline{\quad}$

$11+11+11+11+11+11+11+11+11+11+11$
$= \underline{\quad}$ $11 \times 11 = \underline{\quad}$

Lesson 40

1. Color every 11th square using a light-colored highlighter, crayon or marker.

1	2	3	4	5	6	7	8	9	10
11	12	13	14	15	16	17	18	19	20
21	22	23	24	25	26	27	28	29	30
31	32	33	34	35	36	37	38	39	40
41	42	43	44	45	46	47	48	49	50
51	52	53	54	55	56	57	58	59	60
61	62	63	64	65	66	67	68	69	70
71	72	73	74	75	76	77	78	79	80
81	82	83	84	85	86	87	88	89	90
91	92	93	94	95	96	97	98	99	100

2. Read, whisper and shout (really loud) the colored numbers.

Solve these problems:

$24+10+10+10 = \underline{\qquad}$ $77+10 = \underline{\qquad}$

$13+10+10+10 = \underline{\qquad}$ $11+10 = \underline{\qquad}$

$$
\begin{array}{r} 11 \\ \times\,1 \\ \hline \end{array}
\qquad
\begin{array}{r} 11 \\ \times\,2 \\ \hline \end{array}
\qquad
\begin{array}{r} 11 \\ \times\,3 \\ \hline \end{array}
\qquad
\begin{array}{r} 11 \\ \times\,4 \\ \hline \end{array}
$$

$$
\begin{array}{r} 11 \\ \times\,5 \\ \hline \end{array}
\qquad
\begin{array}{r} 11 \\ \times\,6 \\ \hline \end{array}
\qquad
\begin{array}{r} 11 \\ \times\,7 \\ \hline \end{array}
\qquad
\begin{array}{r} 11 \\ \times\,8 \\ \hline \end{array}
$$

$$
\begin{array}{r} 11 \\ \times\,9 \\ \hline \end{array}
\qquad
\begin{array}{r} 11 \\ \times\,10 \\ \hline \end{array}
\qquad
\begin{array}{r} 11 \\ \times\,11 \\ \hline \end{array}
\qquad
\begin{array}{r} 11 \\ \times\,12 \\ \hline \end{array}
$$

Lesson 41

1. Color every 11th square using a light-colored highlighter, crayon or marker.

1	2	3	4	5	6	7	8	9	10
11	12	13	14	15	16	17	18	19	20
21	22	23	24	25	26	27	28	29	30
31	32	33	34	35	36	37	38	39	40
41	42	43	44	45	46	47	48	49	50
51	52	53	54	55	56	57	58	59	60
61	62	63	64	65	66	67	68	69	70
71	72	73	74	75	76	77	78	79	80
81	82	83	84	85	86	87	88	89	90
91	92	93	94	95	96	97	98	99	100

2. Read, whisper and shout (really loud) the colored numbers.

Solve these problems:

$48+10+10+10 = $ ____ $41+10 = $ ____

$57+10+10+10 = $ ____ $67+10 = $ ____

$$
\begin{array}{cccc}
1 & 2 & 3 & 4 \\
\times\,11 & \times\,11 & \times\,11 & \times\,11 \\
\hline
\end{array}
$$

$$
\begin{array}{cccc}
5 & 6 & 7 & 8 \\
\times\,11 & \times\,11 & \times\,11 & \times\,11 \\
\hline
\end{array}
$$

$$
\begin{array}{cccc}
9 & 10 & 11 & 12 \\
\times\,11 & \times\,11 & \times\,11 & \times\,11 \\
\hline
\end{array}
$$

Lesson 42

1. Color every 12th square using a light-colored highlighter, crayon or marker. These numbers will be 12, 24, 36, 48, 60, 72, 84, 96, 108, 120, 132, and 144.

1	2	3	4	5	6	7	8	9	10
11	12	13	14	15	16	17	18	19	20
21	22	23	24	25	26	27	28	29	30
31	32	33	34	35	36	37	38	39	40
41	42	43	44	45	46	47	48	49	50
51	52	53	54	55	56	57	58	59	60
61	62	63	64	65	66	67	68	69	70
71	72	73	74	75	76	77	78	79	80
81	82	83	84	85	86	87	88	89	90
91	92	93	94	95	96	97	98	99	100

2. Read, whisper and shout (really loud) the colored numbers.

Solve these problems:

$61+10+10+10 = $ _____ $49+10 = $ _____

$34+10+10+10 = $ _____ $94+10 = $ _____

11 × 1	11 × 2	11 × 3	11 × 4
11 × 5	11 × 6	11 × 7	11 × 8
11 × 9	11 × 10	11 × 11	11 × 12

Lesson 43

1. Color every 12th square using a light-colored highlighter, crayon or marker.

1	2	3	4	5	6	7	8	9	10
11	12	13	14	15	16	17	18	19	20
21	22	23	24	25	26	27	28	29	30
31	32	33	34	35	36	37	38	39	40
41	42	43	44	45	46	47	48	49	50
51	52	53	54	55	56	57	58	59	60
61	62	63	64	65	66	67	68	69	70
71	72	73	74	75	76	77	78	79	80
81	82	83	84	85	86	87	88	89	90
91	92	93	94	95	96	97	98	99	100

2. Read, whisper and shout (really loud) the colored numbers.

Solve these problems:

$35+10+10+10 =$ _____ $41+10 =$ _____

$68+10+10+10 =$ _____ $74+10 =$ _____

$1+1+1+1+1+1+1+1+1+1+1+1 =$___ $12 \times 1 =$

$2+2+2+2+2+2+2+2+2+2+2+2 =$___ $12 \times 2 =$

$3+3+3+3+3+3+3+3+3+3+3+3 =$___ $12 \times 3 =$

$4+4+4+4+4+4+4+4+4+4+4+4 =$___ $12 \times 4 =$

$5+5+5+5+5+5+5+5+5+5+5+5 =$___ $12 \times 5 =$

$6+6+6+6+6+6+6+6+6+6+6+6 =$___ $12 \times 6 =$

$7+7+7+7+7+7+7+7+7+7+7+7 =$___ $12 \times 7 =$

$8+8+8+8+8+8+8+8+8+8+8+8 =$___ $12 \times 8 =$

$9+9+9+9+9+9+9+9+9+9+9+9 =$___ $12 \times 9 =$

$10+10+10+10+10+10+10+10+10+10+$
$+10+10 =$___ $12 \times 10 =$ ___

$11+11+11+11+11+11+11+11+11+11+$
$11+ 11 =$___ $12 \times 11 =$ ___

Lesson 44

1. Color every 12th square using a light-colored highlighter, crayon or marker.

1	2	3	4	5	6	7	8	9	10
11	12	13	14	15	16	17	18	19	20
21	22	23	24	25	26	27	28	29	30
31	32	33	34	35	36	37	38	39	40
41	42	43	44	45	46	47	48	49	50
51	52	53	54	55	56	57	58	59	60
61	62	63	64	65	66	67	68	69	70
71	72	73	74	75	76	77	78	79	80
81	82	83	84	85	86	87	88	89	90
91	92	93	94	95	96	97	98	99	100

2. Read, whisper and shout (really loud) the colored numbers.

Solve these problems:

$23+10+10+10 =$ _____ $77+10 =$ _____

$15+10+10+10 =$ _____ $79+10 =$ _____

12	12	12	12
× 1	× 2	× 3	× 4

12	12	12	12
× 5	× 6	× 7	× 8

12	12	12	12
× 9	× 10	× 11	× 12

Lesson 45

1. Color every 12th square using a light-colored highlighter, crayon or marker.

1	2	3	4	5	6	7	8	9	10
11	12	13	14	15	16	17	18	19	20
21	22	23	24	25	26	27	28	29	30
31	32	33	34	35	36	37	38	39	40
41	42	43	44	45	46	47	48	49	50
51	52	53	54	55	56	57	58	59	60
61	62	63	64	65	66	67	68	69	70
71	72	73	74	75	76	77	78	79	80
81	82	83	84	85	86	87	88	89	90
91	92	93	94	95	96	97	98	99	100

2. Read, whisper and shout (really loud) the colored numbers.

Solve these problems:

$33+10+10+10 =$ _____ $77+10 =$ _____

$54+10+10+10 =$ _____ $86+10 =$ _____

$$\begin{array}{r} 1 \\ \times\,12 \\ \hline \end{array} \qquad \begin{array}{r} 2 \\ \times\,12 \\ \hline \end{array} \qquad \begin{array}{r} 3 \\ \times\,12 \\ \hline \end{array} \qquad \begin{array}{r} 4 \\ \times\,12 \\ \hline \end{array}$$

$$\begin{array}{r} 5 \\ \times\,12 \\ \hline \end{array} \qquad \begin{array}{r} 6 \\ \times\,12 \\ \hline \end{array} \qquad \begin{array}{r} 7 \\ \times\,12 \\ \hline \end{array} \qquad \begin{array}{r} 8 \\ \times\,12 \\ \hline \end{array}$$

$$\begin{array}{r} 9 \\ \times\,12 \\ \hline \end{array} \qquad \begin{array}{r} 10 \\ \times\,12 \\ \hline \end{array} \qquad \begin{array}{r} 11 \\ \times\,12 \\ \hline \end{array} \qquad \begin{array}{r} 12 \\ \times\,12 \\ \hline \end{array}$$

Additional Multiplication Problems

Children improve their multiplication speed when they practice multiplying by only one number at a time.

The first multiplication practice set consists of 25 problems per page. There are eleven pages in this practice set. Each page offers your child the opportunity to practice multiplying by only one number, beginning with two and progressing consecutively to twelve. The large print is designed to be friendly and encouraging to your child. Encourage your child to skip count on his/her fingers to obtain the answer.

The second multiplication practice set consists of 50 problems per page. Again, your child is multiplying by only one number. The print size is smaller in this practice set. Do not begin the second practice set until your child has mastered the first practice set.

The third practice set contains 100 mixed problems. Do not allow your child to tackle these problems until he/she has mastered the first two practice sets.

The publisher grants permission to copy these practice sets so that your child may practice the problems over and over again.

Set Number	Page Numbers
1	91-101
2	102-112
3	113-115

$$\begin{array}{r} 2 \\ \times 2 \\ \hline \end{array} \qquad \begin{array}{r} 2 \\ \times 7 \\ \hline \end{array} \qquad \begin{array}{r} 2 \\ \times 4 \\ \hline \end{array} \qquad \begin{array}{r} 2 \\ \times 8 \\ \hline \end{array} \qquad \begin{array}{r} 2 \\ \times 10 \\ \hline \end{array}$$

$$\begin{array}{r} 2 \\ \times 3 \\ \hline \end{array} \qquad \begin{array}{r} 2 \\ \times 5 \\ \hline \end{array} \qquad \begin{array}{r} 2 \\ \times 9 \\ \hline \end{array} \qquad \begin{array}{r} 2 \\ \times 6 \\ \hline \end{array} \qquad \begin{array}{r} 2 \\ \times 11 \\ \hline \end{array}$$

$$\begin{array}{r} 2 \\ \times 7 \\ \hline \end{array} \qquad \begin{array}{r} 2 \\ \times 9 \\ \hline \end{array} \qquad \begin{array}{r} 2 \\ \times 8 \\ \hline \end{array} \qquad \begin{array}{r} 2 \\ \times 2 \\ \hline \end{array} \qquad \begin{array}{r} 2 \\ \times 12 \\ \hline \end{array}$$

$$\begin{array}{r} 11 \\ \times 2 \\ \hline \end{array} \qquad \begin{array}{r} 6 \\ \times 2 \\ \hline \end{array} \qquad \begin{array}{r} 3 \\ \times 2 \\ \hline \end{array} \qquad \begin{array}{r} 4 \\ \times 2 \\ \hline \end{array} \qquad \begin{array}{r} 10 \\ \times 2 \\ \hline \end{array}$$

$$\begin{array}{r} 12 \\ \times 2 \\ \hline \end{array} \qquad \begin{array}{r} 8 \\ \times 2 \\ \hline \end{array} \qquad \begin{array}{r} 5 \\ \times 2 \\ \hline \end{array} \qquad \begin{array}{r} 7 \\ \times 2 \\ \hline \end{array} \qquad \begin{array}{r} 9 \\ \times 2 \\ \hline \end{array}$$

3 × 2	3 × 7	3 × 4	3 × 8	3 ×10
3 × 3	3 × 5	3 × 9	3 × 6	3 ×11
3 × 7	3 × 9	3 × 8	3 × 2	3 ×12
11 × 3	6 × 3	3 × 3	4 × 3	10 × 3
12 × 3	8 × 3	5 × 3	7 × 3	9 × 3

$$\begin{array}{r} 4 \\ \times\,2 \\ \hline \end{array}$$
$$\begin{array}{r} 4 \\ \times\,7 \\ \hline \end{array}$$
$$\begin{array}{r} 4 \\ \times\,4 \\ \hline \end{array}$$
$$\begin{array}{r} 4 \\ \times\,8 \\ \hline \end{array}$$
$$\begin{array}{r} 4 \\ \times\,10 \\ \hline \end{array}$$

$$\begin{array}{r} 4 \\ \times\,3 \\ \hline \end{array}$$
$$\begin{array}{r} 4 \\ \times\,5 \\ \hline \end{array}$$
$$\begin{array}{r} 4 \\ \times\,9 \\ \hline \end{array}$$
$$\begin{array}{r} 4 \\ \times\,6 \\ \hline \end{array}$$
$$\begin{array}{r} 4 \\ \times\,11 \\ \hline \end{array}$$

$$\begin{array}{r} 4 \\ \times\,7 \\ \hline \end{array}$$
$$\begin{array}{r} 4 \\ \times\,9 \\ \hline \end{array}$$
$$\begin{array}{r} 4 \\ \times\,8 \\ \hline \end{array}$$
$$\begin{array}{r} 4 \\ \times\,2 \\ \hline \end{array}$$
$$\begin{array}{r} 4 \\ \times\,12 \\ \hline \end{array}$$

$$\begin{array}{r} 11 \\ \times\,4 \\ \hline \end{array}$$
$$\begin{array}{r} 6 \\ \times\,4 \\ \hline \end{array}$$
$$\begin{array}{r} 3 \\ \times\,4 \\ \hline \end{array}$$
$$\begin{array}{r} 4 \\ \times\,4 \\ \hline \end{array}$$
$$\begin{array}{r} 10 \\ \times\,4 \\ \hline \end{array}$$

$$\begin{array}{r} 12 \\ \times\,4 \\ \hline \end{array}$$
$$\begin{array}{r} 8 \\ \times\,4 \\ \hline \end{array}$$
$$\begin{array}{r} 5 \\ \times\,4 \\ \hline \end{array}$$
$$\begin{array}{r} 7 \\ \times\,4 \\ \hline \end{array}$$
$$\begin{array}{r} 9 \\ \times\,4 \\ \hline \end{array}$$

5 × 2	5 × 7	5 × 4	5 × 8	5 ×10
5 × 3	5 × 5	5 × 9	5 × 6	5 ×11
5 × 7	5 × 9	5 × 8	5 × 2	5 ×12
11 × 5	6 × 5	3 × 5	4 × 5	10 × 5
12 × 5	8 × 5	5 × 5	7 × 5	9 × 5

6 × 2	6 × 7	6 × 4	6 × 8	6 ×10
6 × 3	6 × 5	6 × 9	6 × 6	6 ×11
6 × 7	6 × 9	6 × 8	6 × 2	6 ×12
11 × 6	6 × 6	3 × 6	4 × 6	10 × 6
12 × 6	8 × 6	5 × 6	7 × 6	9 × 6

$$\begin{array}{r} 7 \\ \times\,2 \\ \hline \end{array} \qquad \begin{array}{r} 7 \\ \times\,5 \\ \hline \end{array} \qquad \begin{array}{r} 7 \\ \times\,4 \\ \hline \end{array} \qquad \begin{array}{r} 7 \\ \times\,8 \\ \hline \end{array} \qquad \begin{array}{r} 7 \\ \times\,10 \\ \hline \end{array}$$

$$\begin{array}{r} 7 \\ \times\,3 \\ \hline \end{array} \qquad \begin{array}{r} 7 \\ \times\,7 \\ \hline \end{array} \qquad \begin{array}{r} 7 \\ \times\,9 \\ \hline \end{array} \qquad \begin{array}{r} 7 \\ \times\,6 \\ \hline \end{array} \qquad \begin{array}{r} 7 \\ \times\,11 \\ \hline \end{array}$$

$$\begin{array}{r} 7 \\ \times\,7 \\ \hline \end{array} \qquad \begin{array}{r} 7 \\ \times\,9 \\ \hline \end{array} \qquad \begin{array}{r} 7 \\ \times\,8 \\ \hline \end{array} \qquad \begin{array}{r} 7 \\ \times\,2 \\ \hline \end{array} \qquad \begin{array}{r} 7 \\ \times\,12 \\ \hline \end{array}$$

$$\begin{array}{r} 11 \\ \times\,7 \\ \hline \end{array} \qquad \begin{array}{r} 6 \\ \times\,7 \\ \hline \end{array} \qquad \begin{array}{r} 3 \\ \times\,7 \\ \hline \end{array} \qquad \begin{array}{r} 4 \\ \times\,7 \\ \hline \end{array} \qquad \begin{array}{r} 10 \\ \times\,7 \\ \hline \end{array}$$

$$\begin{array}{r} 12 \\ \times\,7 \\ \hline \end{array} \qquad \begin{array}{r} 8 \\ \times\,7 \\ \hline \end{array} \qquad \begin{array}{r} 5 \\ \times\,7 \\ \hline \end{array} \qquad \begin{array}{r} 7 \\ \times\,7 \\ \hline \end{array} \qquad \begin{array}{r} 9 \\ \times\,7 \\ \hline \end{array}$$

8 × 2	8 × 7	8 × 4	8 × 8	8 ×10
8 × 3	8 × 5	8 × 9	8 × 6	8 ×11
8 × 7	8 × 9	8 × 8	8 × 2	8 ×12
11 × 8	6 × 8	3 × 8	4 × 8	10 × 8
12 × 8	8 × 8	5 × 8	7 × 8	9 × 8

9 × 2	9 × 7	9 × 4	9 × 8	9 ×10
9 × 3	9 × 5	9 × 9	9 × 6	9 ×11
9 × 7	9 × 9	9 × 8	9 × 2	9 ×12
11 × 9	6 × 9	3 × 9	4 × 9	10 × 9
12 × 9	8 × 9	5 × 9	7 × 9	9 × 9

10 × 2	10 × 7	10 × 4	10 × 8	10 × 10
10 × 3	10 × 5	10 × 9	10 × 6	10 × 11
10 × 7	10 × 9	10 × 8	10 × 2	10 × 12
11 ×10	6 ×10	3 ×10	4 ×10	10 ×10
12 ×10	8 ×10	5 ×10	7 ×10	9 ×10

$$\begin{array}{r} 11 \\ \times\,2 \\ \hline \end{array}\qquad \begin{array}{r} 11 \\ \times\,7 \\ \hline \end{array}\qquad \begin{array}{r} 11 \\ \times\,4 \\ \hline \end{array}\qquad \begin{array}{r} 11 \\ \times\,8 \\ \hline \end{array}\qquad \begin{array}{r} 11 \\ \times10 \\ \hline \end{array}$$

$$\begin{array}{r} 11 \\ \times\,3 \\ \hline \end{array}\qquad \begin{array}{r} 11 \\ \times\,5 \\ \hline \end{array}\qquad \begin{array}{r} 11 \\ \times\,9 \\ \hline \end{array}\qquad \begin{array}{r} 11 \\ \times\,6 \\ \hline \end{array}\qquad \begin{array}{r} 11 \\ \times11 \\ \hline \end{array}$$

$$\begin{array}{r} 11 \\ \times\,7 \\ \hline \end{array}\qquad \begin{array}{r} 11 \\ \times\,9 \\ \hline \end{array}\qquad \begin{array}{r} 11 \\ \times\,8 \\ \hline \end{array}\qquad \begin{array}{r} 11 \\ \times\,2 \\ \hline \end{array}\qquad \begin{array}{r} 11 \\ \times12 \\ \hline \end{array}$$

$$\begin{array}{r} 11 \\ \times\,11 \\ \hline \end{array}\qquad \begin{array}{r} 6 \\ \times11 \\ \hline \end{array}\qquad \begin{array}{r} 3 \\ \times11 \\ \hline \end{array}\qquad \begin{array}{r} 4 \\ \times11 \\ \hline \end{array}\qquad \begin{array}{r} 10 \\ \times11 \\ \hline \end{array}$$

$$\begin{array}{r} 12 \\ \times\,11 \\ \hline \end{array}\qquad \begin{array}{r} 8 \\ \times11 \\ \hline \end{array}\qquad \begin{array}{r} 5 \\ \times\,11 \\ \hline \end{array}\qquad \begin{array}{r} 7 \\ \times11 \\ \hline \end{array}\qquad \begin{array}{r} 9 \\ \times\,11 \\ \hline \end{array}$$

12 × 2	12 × 7	12 × 4	12 × 8	12 ×10
12 × 3	12 × 12	12 × 9	12 × 6	12 ×11
12 × 7	12 × 9	12 × 8	12 × 2	12 ×12
11 ×12	6 ×12	3 ×12	4 ×12	10 ×12
12 × 12	8 × 12	12 × 12	7 × 12	9 × 12

2 × 3	2 × 11	2 × 7	2 × 2	2 × 5
2 × 8	2 × 6	2 × 4	2 × 1	2 × 10
2 × 9	2 × 12	2 × 4	2 × 8	2 × 6
2 × 10	2 × 2	2 × 12	2 × 9	2 × 7
2 × 11	2 × 5	2 × 3	2 × 8	2 × 4
2 × 7	2 × 6	3 × 2	8 × 2	6 × 2
12 × 2	1 × 2	7 × 2	5 × 2	11 × 2
2 × 2	6 × 2	9 × 2	5 × 2	4 × 2
7 × 2	3 × 2	8 × 2	2 × 2	10 × 2
9 × 2	7 × 2	6 × 2	8 × 2	4 × 2

$\begin{array}{r} 3 \\ \times\ 3 \\ \hline \end{array}$	$\begin{array}{r} 3 \\ \times\ 11 \\ \hline \end{array}$	$\begin{array}{r} 3 \\ \times\ 7 \\ \hline \end{array}$	$\begin{array}{r} 3 \\ \times\ 2 \\ \hline \end{array}$	$\begin{array}{r} 3 \\ \times\ 5 \\ \hline \end{array}$
$\begin{array}{r} 3 \\ \times\ 8 \\ \hline \end{array}$	$\begin{array}{r} 3 \\ \times\ 6 \\ \hline \end{array}$	$\begin{array}{r} 3 \\ \times\ 4 \\ \hline \end{array}$	$\begin{array}{r} 3 \\ \times\ 1 \\ \hline \end{array}$	$\begin{array}{r} 3 \\ \times\ 10 \\ \hline \end{array}$
$\begin{array}{r} 3 \\ \times\ 9 \\ \hline \end{array}$	$\begin{array}{r} 3 \\ \times\ 12 \\ \hline \end{array}$	$\begin{array}{r} 3 \\ \times\ 4 \\ \hline \end{array}$	$\begin{array}{r} 3 \\ \times\ 8 \\ \hline \end{array}$	$\begin{array}{r} 3 \\ \times\ 6 \\ \hline \end{array}$
$\begin{array}{r} 3 \\ \times\ 10 \\ \hline \end{array}$	$\begin{array}{r} 3 \\ \times\ 2 \\ \hline \end{array}$	$\begin{array}{r} 3 \\ \times\ 12 \\ \hline \end{array}$	$\begin{array}{r} 3 \\ \times\ 9 \\ \hline \end{array}$	$\begin{array}{r} 3 \\ \times\ 7 \\ \hline \end{array}$
$\begin{array}{r} 3 \\ \times\ 11 \\ \hline \end{array}$	$\begin{array}{r} 3 \\ \times\ 5 \\ \hline \end{array}$	$\begin{array}{r} 3 \\ \times\ 3 \\ \hline \end{array}$	$\begin{array}{r} 3 \\ \times\ 8 \\ \hline \end{array}$	$\begin{array}{r} 3 \\ \times\ 4 \\ \hline \end{array}$
$\begin{array}{r} 3 \\ \times\ 7 \\ \hline \end{array}$	$\begin{array}{r} 3 \\ \times\ 6 \\ \hline \end{array}$	$\begin{array}{r} 3 \\ \times\ 3 \\ \hline \end{array}$	$\begin{array}{r} 8 \\ \times\ 3 \\ \hline \end{array}$	$\begin{array}{r} 6 \\ \times\ 3 \\ \hline \end{array}$
$\begin{array}{r} 12 \\ \times\ 3 \\ \hline \end{array}$	$\begin{array}{r} 1 \\ \times\ 3 \\ \hline \end{array}$	$\begin{array}{r} 7 \\ \times\ 3 \\ \hline \end{array}$	$\begin{array}{r} 5 \\ \times\ 3 \\ \hline \end{array}$	$\begin{array}{r} 11 \\ \times\ 3 \\ \hline \end{array}$
$\begin{array}{r} 2 \\ \times\ 3 \\ \hline \end{array}$	$\begin{array}{r} 6 \\ \times\ 3 \\ \hline \end{array}$	$\begin{array}{r} 9 \\ \times\ 3 \\ \hline \end{array}$	$\begin{array}{r} 5 \\ \times\ 3 \\ \hline \end{array}$	$\begin{array}{r} 4 \\ \times\ 3 \\ \hline \end{array}$
$\begin{array}{r} 7 \\ \times\ 3 \\ \hline \end{array}$	$\begin{array}{r} 3 \\ \times\ 3 \\ \hline \end{array}$	$\begin{array}{r} 8 \\ \times\ 3 \\ \hline \end{array}$	$\begin{array}{r} 2 \\ \times\ 3 \\ \hline \end{array}$	$\begin{array}{r} 10 \\ \times\ 3 \\ \hline \end{array}$
$\begin{array}{r} 9 \\ \times\ 3 \\ \hline \end{array}$	$\begin{array}{r} 7 \\ \times\ 3 \\ \hline \end{array}$	$\begin{array}{r} 6 \\ \times\ 3 \\ \hline \end{array}$	$\begin{array}{r} 8 \\ \times\ 3 \\ \hline \end{array}$	$\begin{array}{r} 4 \\ \times\ 3 \\ \hline \end{array}$

4 × 3	4 × 11	4 × 7	4 × 2	4 × 5
4 × 8	4 × 6	4 × 4	4 × 1	4 × 10
4 × 9	4 × 12	4 × 4	4 × 8	4 × 6
4 × 10	4 × 2	4 × 12	4 × 9	4 × 7
4 × 11	4 × 5	4 × 3	4 × 8	4 × 4
4 × 7	4 × 6	3 × 4	8 × 4	6 × 4
12 × 4	1 × 4	7 × 4	5 × 4	11 × 4
2 × 4	6 × 4	9 × 4	5 × 4	4 × 4
7 × 4	3 × 4	8 × 4	2 × 4	10 × 4
9 × 4	7 × 4	6 × 4	8 × 4	4 × 4

5 × 3	5 × 11	5 × 7	5 × 2	5 × 5
5 × 8	5 × 6	5 × 4	5 × 1	5 × 10
5 × 9	5 × 12	5 × 4	5 × 8	5 × 6
5 × 10	5 × 2	5 × 12	5 × 9	5 × 7
5 × 11	5 × 5	5 × 3	5 × 8	5 × 4
5 × 7	5 × 6	3 × 5	8 × 5	6 × 5
12 × 5	1 × 5	7 × 5	5 × 5	11 × 5
2 × 5	6 × 5	9 × 5	5 × 5	4 × 5
7 × 5	3 × 5	8 × 5	2 × 5	10 × 5
9 × 5	7 × 5	6 × 5	8 × 5	4 × 5

6 × 3	6 × 11	6 × 7	6 × 2	6 × 5
6 × 8	6 × 6	6 × 4	6 × 1	6 × 10
6 × 9	6 × 12	6 × 4	6 × 8	6 × 6
6 × 10	6 × 2	6 × 12	6 × 9	6 × 7
6 × 11	6 × 5	6 × 3	6 × 8	6 × 4
6 × 7	6 × 6	3 × 6	8 × 6	6 × 6
12 × 6	1 × 6	7 × 6	5 × 6	11 × 6
2 × 6	6 × 6	9 × 6	5 × 6	4 × 6
7 × 6	3 × 6	8 × 6	2 × 6	10 × 6
9 × 6	7 × 6	6 × 6	8 × 6	4 × 6

7 × 3	7 × 11	7 × 7	7 × 2	7 × 6
7 × 8	7 × 6	7 × 4	7 × 1	7 × 10
7 × 9	7 × 12	7 × 4	7 × 8	7 × 6
7 × 10	7 × 2	7 × 12	7 × 9	7 × 7
7 × 11	7 × 6	7 × 3	7 × 8	7 × 4
7 × 7	7 × 6	3 × 7	8 × 7	6 × 7
12 × 7	1 × 7	7 × 7	6 × 7	11 × 7
2 × 7	6 × 7	9 × 7	6 × 7	4 × 7
7 × 7	3 × 7	8 × 7	2 × 7	10 × 7
9 × 7	7 × 7	6 × 7	8 × 7	4 × 7

8 × 3	8 × 11	8 × 7	8 × 2	8 × 7
8 × 8	8 × 6	8 × 4	8 × 1	8 × 10
8 × 9	8 × 12	8 × 4	8 × 8	8 × 6
8 × 10	8 × 2	8 × 12	8 × 9	8 × 7
8 × 11	8 × 7	8 × 3	8 × 8	8 × 4
8 × 7	8 × 6	3 × 8	8 × 8	6 × 8
12 × 8	1 × 8	7 × 8	7 × 8	11 × 8
2 × 8	6 × 8	9 × 8	7 × 8	4 × 8
7 × 8	3 × 8	8 × 8	2 × 8	10 × 8
9 × 8	7 × 8	6 × 8	8 × 8	4 × 8

9 × 3	9 × 11	9 × 7	9 × 2	9 × 8
9 × 8	9 × 6	9 × 4	9 × 1	9 × 10
9 × 9	9 × 12	9 × 4	9 × 8	9 × 6
9 × 10	9 × 2	9 × 12	9 × 9	9 × 7
9 × 11	9 × 8	9 × 3	9 × 8	9 × 4
9 × 7	9 × 6	3 × 9	8 × 9	6 × 9
12 × 9	1 × 9	7 × 9	8 × 9	11 × 9
2 × 9	6 × 9	9 × 9	8 × 9	4 × 9
7 × 9	3 × 9	8 × 9	2 × 9	10 × 9
9 × 9	7 × 9	6 × 9	8 × 9	4 × 9

10 × 3	10 × 11	10 × 7	10 × 2	10 × 8
10 × 8	10 × 6	10 × 4	10 × 1	10 × 10
10 × 9	10 × 12	10 × 4	10 × 8	10 × 6
10 × 10	10 × 2	10 × 12	10 × 9	10 × 7
10 × 11	10 × 8	10 × 3	10 × 8	10 × 4
10 × 7	10 × 6	3 × 10	8 × 10	6 × 10
12 × 10	1 × 10	7 × 10	8 × 10	11 × 10
2 × 10	6 × 10	9 × 10	8 × 10	4 × 10
7 × 10	3 × 10	8 × 10	2 × 10	10 × 10
9 × 10	7 × 10	6 × 10	8 × 10	4 × 10

11 × 3	11 × 11	11 × 7	11 × 2	11 × 8
11 × 8	11 × 6	11 × 4	11 × 1	11 × 10
11 × 9	11 × 12	11 × 4	11 × 8	11 × 6
11 × 10	11 × 2	11 × 12	11 × 9	11 × 7
11 × 11	11 × 8	11 × 3	11 × 8	11 × 4
11 × 7	11 × 6	3 × 11	8 × 11	6 × 11
12 × 11	1 × 11	7 × 11	8 × 11	11 × 11
2 × 11	6 × 11	9 × 11	8 × 11	4 × 11
7 × 11	3 × 11	8 × 11	2 × 11	10 × 11
9 × 11	7 × 11	6 × 11	8 × 11	4 × 11

12 × 3	12 × 11	12 × 7	12 × 2	12 × 8
12 × 8	12 × 6	12 × 4	12 × 1	12 × 10
12 × 9	12 × 8	12 × 4	12 × 8	12 × 6
12 × 10	12 × 2	12 × 12	12 × 9	12 × 7
12 × 11	12 × 12	12 × 3	12 × 8	12 × 4
12 × 7	12 × 6	3 × 12	8 × 12	6 × 12
12 × 12	1 × 12	7 × 12	8 × 12	11 × 12
2 × 12	6 × 12	9 × 12	8 × 12	4 × 12
7 × 12	3 × 12	8 × 12	2 × 12	10 × 12
9 × 12	7 × 12	6 × 12	8 × 12	4 × 12

2 × 3	2 × 5	4 × 7	6 × 8	2 × 12	3 × 9	5 × 11	8 × 8	2 × 6	11 × 5
6 × 11	12 × 11	10 × 4	1 × 7	7 × 9	6 × 4	12 × 8	1 × 10	4 × 10	8 × 6
12 × 6	4 × 6	9 × 6	8 × 11	6 × 9	7 × 5	10 × 9	11 × 12	12 × 4	4 × 11
9 × 3	2 × 10	2 × 7	11 × 8	5 × 7	9 × 10	7 × 6	1 × 4	9 × 12	4 × 12
10 × 10	8 × 4	3 × 8	3 × 5	11 × 9	9 × 5	11 × 6	9 × 9	2 × 9	8 × 3
7 × 8	9 × 4	6 × 12	7 × 7	9 × 7	1 × 12	12 × 7	8 × 9	4 × 5	11 × 4
1 × 8	2 × 4	5 × 6	6 × 6	10 × 8	4 × 3	4 × 9	5 × 8	8 × 10	6 × 5
1 × 5	7 × 12	12 × 12	5 × 5	1 × 3	9 × 8	6 × 10	4 × 8	6 × 5	10 × 12
7 × 11	1 × 9	5 × 4	5 × 3	3 × 11	8 × 7	3 × 3	10 × 6	6 × 7	5 × 5
12 × 10	6 × 3	10 × 3	8 × 5	10 × 5	11 × 7	12 × 3	3 × 7	1 × 6	5 × 10

9 × 2	10 × 3	8 × 7	6 × 3	1 × 4	3 × 6	5 × 6	8 × 8	12 × 4	4 × 5
6 × 7	7 × 11	3 × 4	1 × 3	8 × 5	6 × 4	12 × 8	1 × 9	1 × 1	8 × 6
12 × 7	3 × 5	7 × 2	2 × 11	3 × 7	7 × 6	3 × 3	11 × 12	12 × 4	4 × 5
5 × 3	5 × 10	2 × 7	11 × 6	5 × 5	8 × 2	7 × 6	1 × 4	9 × 12	10 × 5
5 × 7	6 × 1	10 × 2	7 × 5	11 × 9	9 × 5	4 × 6	9 × 6	2 × 2	12 × 5
7 × 8	9 × 4	6 × 12	1 × 7	9 × 7	1 × 3	12 × 1	8 × 9	4 × 2	11 × 1
1 × 6	2 × 12	5 × 6	6 × 6	10 × 8	4 × 7	3 × 9	12 × 3	8 × 10	11 × 3
1 × 8	7 × 4	1 × 2	5 × 2	1 × 3	9 × 8	6 × 6	4 × 3	6 × 2	10 × 12
7 × 7	1 × 2	1 × 4	5 × 3	3 × 11	8 × 7	3 × 3	10 × 7	6 × 2	9 × 5
7 × 10	6 × 3	4 × 4	9 × 9	10 × 5	10 × 2	12 × 3	3 × 7	7 × 6	5 × 9

4 ×2	10 ×5	8 ×7	6 ×3	3 ×4	3 ×6	5 ×6	8 ×7	12 ×4	4 ×2
6 ×7	7 ×12	3 ×4	1 ×3	9 ×5	6 ×4	11 ×8	1 ×9	2 ×1	8 ×7
10 ×7	6 ×4	7 ×2	3 ×11	3 ×7	7 ×6	3 ×3	11 ×10	12 ×4	4 ×5
8 ×3	5 ×11	2 ×7	12 ×6	5 ×6	9 ×2	7 ×6	1 ×6	9 ×12	10 ×5
7 ×7	8 ×5	10 ×3	7 ×5	11 ×9	9 ×9	4 ×6	9 ×6	2 ×2	12 ×6
7 ×8	9 ×4	6 ×12	1 ×7	9 ×9	1 ×3	12 ×1	8 ×9	4 ×2	11 ×2
1 ×6	3 ×12	5 ×6	6 ×6	12 ×8	4 ×7	3 ×9	12 ×3	8 ×10	11 ×3
3 ×8	7 ×4	1 ×2	5 ×2	1 ×3	9 ×8	6 ×6	4 ×3	6 ×2	11 ×12
9 ×7	1 ×3	4 ×4	5 ×5	3 ×11	8 ×7	3 ×3	10 ×7	6 ×2	9 ×5
8 ×10	6 ×4	4 ×4	9 ×9	11 ×5	10 ×2	12 ×8	4 ×7	7 ×5	5 ×9

Additional Ideas to Practice Multiplication

Multiplication War

A minimum of two players is necessary. There is no maximum.

The game is played with a standard deck of cards. Remove the four jacks, four queens, four kings, and jokers.

Deal out the remaining cards to each player evenly.

Each player turns over two cards and multiplies them. The player with the highest product wins all the cards. The **product** is the answer when two numbers are multiplied together.

If the product is the same for any two or more players, then these players have a "war." Each of these players put down three cards face-down. They then put down two more cards, face-up. The face-up cards are multiplied. If the product is again the same, the "war" is repeated. The winner is the one with the highest product. The winner takes all the cards.

Continue until one player has all the cards.

"I Am Thinking of Two Numbers"

A minimum of two players is necessary. There is no maximum.

One player thinks of two numbers which he/she adds together. He/she then multiplies the numbers. He/she then tells the other player(s) the sum and the product of the two numbers. The **sum** is the answer when two numbers are added together. The **product** is the answer when two numbers are multiplied together. The other player(s) must guess the original two numbers.

For example: First player chooses 5 and 3. He/she announces "I am thinking of two numbers that sum to 8 and have a product of 15." The other player(s) must figure out that the original numbers were 5 and 3.